BIOFILTRATION
FOR
AIR POLLUTION
CONTROL

BIOFILTRATION
FOR
AIR POLLUTION
CONTROL

Joseph S. Devinny
Marc A. Deshusses
Todd S. Webster

CRC Press
Taylor & Francis Group
Boca Raton London New York

CRC Press is an imprint of the
Taylor & Francis Group, an **informa** business

First published in 1999 by Lewis Publishers

Published in 2020 by CRC Press
Taylor & Francis Group
6000 Broken Sound Parkway NW, Suite 300
Boca Raton, FL 3487-2742

First issued in paperback 2020

© 1999 by Taylor & Francis Group, LLC
CRC Press is an imprint of Taylor & Francis Group, an Informa business

No claim to original U.S. Government works

ISBN-13: 978-0-367-57925-8 (pbk)
ISBN-13: 978-1-56670-289-8 (hbk)

Visit the Taylor & Francis Web site at
http://www.taylorandfrancis.com

and the CRC Press Web site at
http://www.crcpress.com

Library of Congress Cataloging-in-Publication Data

Catalog record is available from the Library of Congress

Preface

Despite great efforts and expenditures, air pollution remains a pressing environmental problem and the number one environmental threat to public health. Air pollution sources are numerous. They are diverse in contaminant type, concentration, and amount of discharge. The industries and other activities that produce air pollution vary in the processes they use, in the difficulty of controlling their discharges, and in their ability to pay for remediation. Successful air pollution control will require equally diverse technologies, capable of handling large and small discharges of every description. Most important and perhaps most difficult, each pollution source must be provided with a treatment technology that is economically supportable by the activity it serves.

Biofiltration will certainly be one of the technologies employed. It is an established approach in Europe and is increasingly being applied in North America. Ideally, it converts pollutants to harmless products, requires no fuel, utilizes inexpensive off-the-shelf components, generates no hazardous by-products, and is ultimately a low-cost alternative for appropriate applications. The concept is fundamentally straightforward: contaminants can be collected on a damp porous medium and biodegraded. But, the physical, chemical, and ecological rules that apply combine to make a complex and only partially understood system. Biofiltration is simple and inexpensive when it is done right, but the knowledge needed to "do it right" is at the frontiers of science and engineering.

This book is intended to provide the reader with a thorough introduction to that knowledge. It covers the concepts that engineers, consultants, researchers, policymakers, and general scientists must use to design and operate biofilters. Though numerous papers have addressed biofiltration, this is the first book attempting a comprehensive survey. It introduces the general scientific principles of the technology while providing numerous references for those seeking more detailed information.

Biofiltration for Air Pollution Control includes chapters on mechanisms of biofiltration, biofilter media, factors influencing and controlling biofiltration, microbial ecology in biofilters, biofilter modeling, design and costs, and startup monitoring. Some biofilter case studies are provided. It is the authors' hope that it will be a valuable starting point for those who wish to harness the power of microorganisms in the effort to clean the air.

We wish to acknowledge our many colleagues who provided generous assistance in improving the content of this book and are indebted to those individuals and companies who generously provided permission to reprint illustrative material. These include B. Baltzis (New Jersey Institute of Technology), H. Bohn (Bohn Biofilter Corp.), H. Cox (University of California, Riverside), R. Fuller (U.S. Filter), W. Fucich (Envirogen, Inc.), R. Heuckeroth (AMETEK-Rotron Biofiltration Products), D. Jacobson (Envirogen, Inc.), S. Kampeter (Monsanto Enviro-Chem), G. Leson (Leson Environmental Consulting), J. Loy (Zander Umwelt GmbH), Mike McGrath (Monsanto Enviro-Chem), H.-J. Mildenberger (Novartis Services AG), D. Nisi (Mercedes-Benz AG), S. Ottengraf (Eindhoven University of Technology, The Netherlands), P. Petro (Environmental Resolutions, Inc.), K. Romstad (Environmental Resolutions, Inc.), G. Sears (KPMG Management), E. Schroeder (University of California, Davis), J. Scillieri (Applied Air Systems, Inc.), Z. Shareefdeen (King Fahd University of Petroleum and Minerals), G. Skladany (Fluor Daniel GTI, Inc.), S. Standefer (PPC Biofilter), B. Stewart (Bio-Reaction Industries, Inc.), P. Togna (Envirogen, Inc.), C. van Lith (ClairTech), N. Webster (Webster Environmental Associates, Inc.), A. White (Microbial Insights, Inc.), and R. Willingham (PPC Biofilter).

The authors

Joseph S. Devinny, Ph.D., is Professor of Civil and Environmental Engineering and Associate Dean for Academic Affairs at the University of Southern California. He earned a B.S. in Chemistry at the California Institute of Technology, an M.S. in Chemistry at the University of Oregon, and his Ph.D. in Environmental Engineering at the California Institute of Technology. His research and teaching have been related to biological aspects of Environmental Engineering, including preservation of natural ecosystems, biological treatment of contaminated soils, and biofiltration of air.

Marc A. Deshusses, Ph.D., is an Assistant Professor of Chemical Engineering and a faculty member in the Microbiology Graduate Program and in the Environmental Toxicology Program at the University of California, Riverside. He received his B.S., M.S., and Ph.D. degrees in Chemical Engineering from the Swiss Federal Institute of Technology. He is author of more than 40 scientific papers, reports, and conference proceedings dealing with biological techniques for air pollution control and biodegradation or organics contaminants. His research interest is the development and application of systems for bioremediation of organic wastes in air, water, and soils.

Todd S. Webster, Ph.D., is a project manager/group leader for Envirogen, Inc.'s (Lawrenceville, NJ) California-based biological remediation projects. He obtained his Masters and Ph.D. in Environmental Engineering from the University of Southern California, Los Angeles. He has directed numerous laboratory, pilot-scale, and field investigations of both biofiltration and biotrickling filtration systems. His research interests include biological treatment systems for air, water, and soil application, as well as problems associated with natural ecosystem management.

Contents

Dedication

We dedicate this book to our wives, Betty Bluml, Sandra Deshusses, and Sandra Webster, whose patience and support helped make it possible.

chapter one

Introduction

Within the past 40 years, the medical and scientific communities have begun to comprehend the deleterious chronic effects of air pollution. Whether it is volatile organic compounds (VOCs) acting as catalysts for smog formation, chlorinated compounds depleting the ozone layer, or odorous compounds becoming a human nuisance, the need to control and treat air pollution has become an urgent environmental and medical concern. The urgency of the matter is clearly evident in the numerous global, national, and regional environmental air pollution control regulations developed to maintain healthful air quality. Such stringent regulations have driven both the industrial and commercial sectors to rely on the technologies of carbon adsorption, incineration, or scrubbing to lessen the environmental damage caused by technological advances. However, as regulations are further refined to control air emissions under stricter standards, such air pollution control technologies will become more costly. As health risks for additional air phase contaminants are discovered, regulations restricting the emissions of these contaminants will also be introduced. For these reasons, industrial and commercial sectors that have had limited need for air pollution control in the past will suddenly find themselves immersed in regulatory compliance activities.

Since the early twentieth century, biological treatment processes have found wide application in wastewater and solid waste pollution control. The need for alternative cost-effective waste gas treatment technologies has led to similar biological treatment processes for waste gas streams. One such treatment technology is biofiltration. In biofiltration, a humid, contaminated air stream is passed through a porous support material on which pollutant-degrading cultures are immobilized. Like most biological treatment processes, biofiltration relies on microbial catabolic reactions for the degradation of waste compounds. Biofilters have found most of their success in the treatment of dilute, high-flow waste gas streams containing odors or volatile organic compounds. Under optimal conditions, the pollutants can be degraded completely to carbon dioxide, water, and excess biomass. Such a system holds promise to treat many of the same contaminants that have been handled by wastewater treatment plants over the past 100 years. However, like all emerging technologies, biofiltration has an appropriate niche in the industrial and commercial sectors. It cannot be considered a panacea for all operations and every industry.

This first chapter serves as an introduction to biofiltration, explaining the forces that have led to its development and further advancement in the waste gas treatment market. In order to assist the reader in understanding the focus of the book, air pollution control regulatory issues, the alternative control technologies available, a brief history of biofiltration, the current marketplace for biofiltration, common biofilter terminology, and additional resources are described in the first chapter.

1.1   Air pollution legislation

As the effects of air pollution on all life forms have become better understood, environmental legislation controlling the emission of volatile organic compounds, toxics, and odors has proliferated. Enforcement of such regulations by federal, state, and regional government agencies has forced industry to comply. In general, the role of federal agencies is to establish baseline emission standards. These standards reflect potential health risks (cancer, respiratory damage, etc.) and environmental degradation (smog precursors, greenhouse gas effects, ozone depletion, acid rain, etc.) that the contaminant emissions create. The state and local governments may establish stricter standards as needed to ease the effects of contamination on more populated areas of the country. Such a trickle-down effect of regulation provides adequate enforcement from a national to a local level and prevents some companies from avoiding participation in air pollution control while others are forced to comply. Though regulations differ from country to country, all nations are finding a growing need for air quality control.

In the U.S., the enactment of the federal Clean Air Act Amendments of 1990 has brought about a strict regulation of air emissions. The regulation, under section Title III, calls for 189 chemicals to be considered air toxics. A 90% reduction in the production of these contaminants is required by the year 2000 (Zahodiakin, 1995). Any facility which emits at least 10 tons per year of a listed pollutant or a total of 25 tons of listed pollutants will be required to install "maximum achievable control technologies" (MACT). The measure also calls for a 15% reduction in ground-level ozone for the country's most polluted areas and a phase-out of chlorinated fluorocarbons (CFCs), carbon tetrachloride, and hydrochlorofluorocarbons (HCFCs).

Though not all countries use the same emission standards, the trend towards stricter regulation of air emissions is general throughout the world. As groups such as the World Health Organization (WHO) chronicle the harmful effects of chemicals on human health, the public and their respective governments will continue to take action to prevent exposure to these harmful contaminants. As technology and industry throughout the world advance, complete elimination of such chemicals seems impossible. In the future, effective air contaminant reduction through "end of the pipe" control technologies will be used by companies to achieve economic growth while minimizing environmental contamination.

1.2 Types of waste gas treatment

There are two forms of applicable air emissions control. Source control involves the reduction of emissions through raw product substitution, reduction, or recycling. However, these reduction mechanisms may reduce the quality of the product or may increase costs. Secondary control involves treatment of the waste gas after it has been produced. The choice of technology is often dictated by economic and ecological constraints. Such constraints arise from the nature of the compound being treated, the concentration, the flow rate, and the mode of emission of the gaseous waste stream. Combinations of various technologies may often be required to meet regulatory standards.

1.2.1 Condensation

Waste gas contaminants that are concentrated and have a high boiling point may be partially recovered by simultaneous cooling and compressing of the gaseous vapors. This technique is only economical for concentrated vapors where there is some recycle or recovery value. If the waste gas is a mixed pollutant stream, recycling will be virtually impossible, and incineration of the condensed liquid may be required. This technique must often be followed by additional removal technologies for compliance with regulatory emission standards.

1.2.2 Incineration

Thermal and catalytic incineration are widely used and effective treatment processes for waste gases. Thermal incineration involves the combustion of pollutants at temperatures of 700 to 1400°C. Catalytic incineration allows process temperatures between 300 and 700°C with catalysts such as platinum, palladium, and rubidium. Incineration is the most widely used secondary technique, but costs are high for low-concentration pollutant vapors because of the need for large amounts of fuel. Regenerative or recuperative heat systems are often used as an attempt to reduce these fuel-operating costs. Production of nitrogen oxides (NO_x) and some dioxins is also possible during incineration. In general, the technology is more suitable for concentrated streams with moderate flow rates.

1.2.3   Adsorption

Adsorption generally occurs on a fixed or fluidized bed of material such as activated carbon or zeolite and is most efficient for treatment of low concentration vapors. The effectiveness of a carbon adsorption system for a particular waste stream is a function of the air flow rate, the total VOC loading of the stream, and the individual components of the VOC stream. Adsorption is generally used for controlling VOCs with low vapor pressures and high molecular weights. Once the activated bed has reached adsorptive capacity, the material must be removed and often treated as hazardous waste, increasing the operating costs of the system. Difficulties in assessing the exact point of complete bed exhaustion are also often encountered, and the bed may be renewed before it is necessary. Such misjudgments drive operating costs upward. The regeneration of the carbon is possible with pollutant recovery by desorption with steam or hot air. However, disposal or incineration of the spent carbon is often more economical.

1.2.4   Absorption

Absorption removes the waste gas contaminant with a scrubbing solution. The gas enters a large contactor where the gaseous pollutants are transferred to a liquid phase. Efficient gas-liquid mass transfer may be accomplished through the use of a packed or bubble column, or a venturi contactor. Success is dictated by the affinity of the pollutant for the liquid phase. Water is the most frequently used scrubbing solution, and the pH can be adjusted to increase the solubility of acidic or basic gases. For hydrophobic pollutants, organic solvents such as silicon oil may also be used as scrubbing solutions. Once the pollutant transfer has occurred, additional treatment of the liquid phase may be necessary. This may be achieved by desorbing the pollutant at high temperatures and incinerating the vapors. If the scrubbing solution is water, the wastewater may be directed to a treatment plant.

1.2.5   Membrane systems

Membrane systems can be used to transfer VOCs from an air stream to a water phase. In a membrane separation system, compression and condensation of the emission stream is followed by membrane separation. By compressing the emission stream to approximately 310 to 1400 kPa, a higher vapor pressure can be maintained on the air-feed side than on the permeate side of the membrane. This pressure differential drives the membrane separation process. The compressed mixture can be processed through a condenser where portions of the organic vapors are recovered. The remaining air stream is then passed across the surface of a microporous hydrophobic membrane constructed of materials such as polyethylene and polypropylene. The membrane provides high gas permeability without allowing bulk permeate transport across the membrane. The membrane pores remain filled with water, and the organic vapors are transferred through the membrane by the differential pressure. The resulting products are a permeate stream containing the majority of the organic compounds and an air stream containing residual organic compounds. The process then requires treatment of the liquid permeate for final VOC disposal or recycling.

1.2.6 Biological treatment

Gas-phase biological reactors utilize microbial metabolic reactions to treat contaminated air. Biological treatment is effective and economical for low concentrations of contaminant in large quantities of air. The contaminants are sorbed from a gas to an aqueous phase where microbial attack occurs. Through oxidative and occasionally reductive reactions, the contaminants are converted to carbon dioxide, water vapor, and organic biomass. These air pollutants may be either organic or inorganic vapors and are used as energy and sometimes as a carbon source for maintenance and growth by the microorganism populations. In general, the microbes used for biological treatment are organisms that are naturally occurring. These microbial populations may be dominated by one particular microbial species or may interact with numerous species to attack a particular type of contaminant synergistically . Microbes within these biological treatment systems are also engaged in many of the same ecological relationships (predation, parasitism, etc.) that are typical to macroorganisms. Such relationships are necessary to provide an important balance within the system.

The particular contaminants of interest must be biodegradable and non-toxic for biological air treatment to be successful. The most successful removal in gas-phase bioreactors occurs for low molecular weight and highly soluble organic compounds with simple bond structures. Compounds with complex bond structures generally require more energy to be degraded, and this energy is not always available to the microbes. Hence, little or no biodegradation of these types of compounds occurs. Instead, microorganisms degrade those compounds that are readily available and easier to degrade. Organic compounds such as alcohols, aldehydes, ketones, and some simple aromatics demonstrate excellent biodegradability. Some compounds that show moderate to slow degradation include phenols, chlorinated hydrocarbons, polyaromatic hydrocarbons, and highly halogenated hydrocarbons. Inorganic compounds such as hydrogen sulfide and ammonia are also biodegraded well. Certain anthropogenic compounds may not biodegrade at all because microorganisms do not possess the necessary enzymes to break the bond structure of the compound effectively (Table 1.1). Compounds that are biologically treatable can come from a wide array of sources. Some industrial processes where biological air pollution control may be effectively utilized are listed in Table 1.2.

Table 1.1 Biodegradability of Various Contaminants in a Biofilter

Contaminant	Biodegradability	Contaminant	Biodegradability
Aliphatic hydrocarbons		*Sulfur-containing[b] carbon compounds*	
Methane	1	Carbon disulfide	2
Propane	?	Dimethyl sulfide	2
Butane	?	Dimethyl disulfide	2
Pentane	1	Methyl mercaptan	1
Isopentane	1	Thiocyanates	1
Hexane	2		
Cyclohexane	1	*Oxygenated carbon compounds*	
Acetylene	1	**Alcohols**	3
		Methanol	3
Aromatic hydrocarbons		Ethanol	3
Benzene	2	Butanol	3
Phenol	3	2-Butanol	3
Toluene	3	1-Propanol	3
Xylene	2	2-Propanol	3
Styrene	2	**Aldehydes**	3
Ethylbenzene	3	Formaldehyde	3
		Acetaldehyde	3
Chlorinated[b] hydrocarbons		**Carbonic acids (esters)**	3
Carbon tetrachloride	1	Butyric acid	3
Chloroform	1	Vinyl acetate	2
Dichloromethane	3	Ethyl acetate	3
Bromodichloromethane	?	Butyl acetate	3
1,1,1-Trichloroethane	?	Isobutyl acetate	3
1,1-Dichloroethane	?	**Ethers**	1
Tetrachloroethene	1[a]	Diethyl ether	1
Trichloroethene	1[a]	Dioxane	1
1,2-Dichloroethane	?	Methyl tert-butyl ether	1
1,1-Dichloroethene	?	Tetrahydrofuran	3
Vinyl chloride	1	**Ketones**	3
1,2-Dichlorobenzene	?	Acetone	3
Chlorotoluene	1	Methyl ethyl ketone	3
		Methyl isobutyl ketone	3
Nitrogen-containing carbon compounds			
Amines	3	*Inorganic[b] compounds*	
Aniline	3	Ammonia	3
Nitriles	1	Hydrogen sulfide	3
Acrylonitrile	?	Nitrogen oxide	1
Pyridine	1		

[a] Indicates that cometabolism or anaerobic treatment has been identified within a biofilter.

[b] Indicates that a change in filter bed pH may occur with treatment of these compounds. This change may negatively affect performance.

Note: 1 = some biodegradability; 2 = moderate biodegradability; 3 = good biodegradability; ? = unknown.

Table 1.2   Industries Where Biological Air Pollution Control May Be Appropriate

Adhesive production	Food processing	Petroleum industry
Animal husbandry	Fragrance industry	Printing industry
Chemical manufacturing	Furniture manufacturing	Pulp and paper
Chemical storage	Investment foundries	Rendering
Coatings industry	Iron foundries	Sewage treatment
Composting	Landfill gas extraction	Soil-vapor extraction
Crematorium	Petrochemical manufacturing	Wood products production

Though a number of different configurations exist, the major air-phase biological reactors are biofilters, biotrickling filters, and bioscrubbers. The basic removal mechanisms are similar for all reactor types; however, differences exist in the phase of the microbes, which may be suspended or fixed, and the state of the liquid, which may be flowing or stationary (Table 1.3).

1.2.6.1   Biofilters

Biofiltration uses microorganisms fixed to a porous medium to break down pollutants present in an air stream. The microorganisms grow in a biofilm on the surface of a medium or are suspended in the water phase surrounding the medium particles (Figure 1.1). The filter-bed medium consists of relatively inert substances (compost, peat, etc.) which ensure large surface attachment areas and additional nutrient supply. As the air passes through the bed, the contaminants in the air phase sorb into the biofilm and onto the filter medium, where they are biodegraded (Figure 1.2). Biofilters are not filtration units as strictly defined. Instead, they are systems that use a combination of basic processes: absorption, adsorption, degradation, and desorption of gas-phase contaminants.

Biofilters usually incorporate some form of water addition to control moisture content and add nutrients. In general, the gas stream is humidified before entering the biofilter reactor. However, if humidification proves inadequate, direct irrigation of the bed may be needed.

The overall effectiveness of a biofilter is largely governed by the properties and characteristics of the support medium, which include porosity, degree of compaction, water retention capabilities, and the ability to host microbial populations. Critical biofilter operational and performance parameters include

Table 1.3   Classification of Bioreactors
for Waste Gas Purification

Reactor type	Microorganisms	Water phase
Biofilter	Fixed	Stationary
Biotrickling filter	Fixed	Flowing
Bioscrubber	Suspended	Flowing

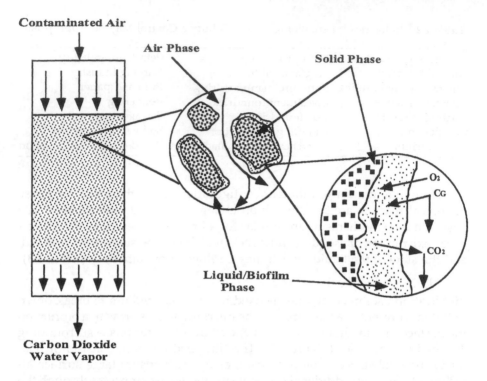

Figure 1.1 Internal mechanisms of a biofilter. Contaminated air (C_G) passes through the filter bed medium (compost, peat, soil, etc.) with oxygen (O_2) and sorbs into a microbial biofilm/liquid phase attached to the filter medium. Microbes convert the contaminant to carbon dioxide (CO_2) and water.

the microbial inoculum, medium pH, temperature, and the medium moisture and nutrient content (Table 1.4).

1.2.6.2   Biotrickling filters and bioscrubbers

In biotrickling filters and bioscrubbers, gas contaminants are absorbed in a free liquid phase prior to biodegradation by either suspended or immobilized microorganisms (Figure 1.3). For biotrickling filters, microbes fixed to an inorganic packing material and suspended microbes in the water phase degrade the absorbed contaminants as they pass through the reactor. Biotrickling filters operate with the air and water phases moving either counter-currently or co-currently, depending on the specific operation. As the water is recirculated, nutrients, acids, or bases may be added by the operator to regulate the environment for optimal pollutant removal. Biotrickling filters are governed by many of the same phenomena as biofilters. Most importantly, a biotrickling filter reactor must host a thriving microbiological population while avoiding conditions that promote excessive biomass growth and clogging conditions.

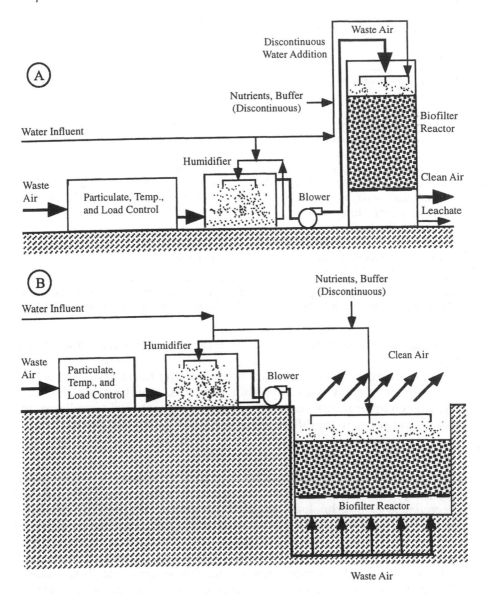

Figure 1.2 Schematic diagrams of **(A)** above-ground closed biofilter and **(B)** below-ground open biofilter.

In a bioscrubber, after initial contaminant absorption occurs, the degradation of the contaminants is performed by a suspended consortium of microbes in a separate vessel. Absorption may be achieved in a packed column, spray tower, or a bubble column. The water is transferred to a separate vessel where optimal environmental conditions for degradation

Table 1.4   Typical Biofilter Operating Conditions for Waste Air Treatment

Parameter	Typical value
Biofilter layer height	1–1.5 m
Biofilter area	1–3000 m²
Waste air flow	50–300,000 m³ h⁻¹
Biofilter surface loading	5–500 m³ m⁻² h⁻¹
Biofilter volumetric loading	5–500 m³ m⁻³ h⁻¹
Bed void volume	50%
Mean effective gas residence time	15–60 s
Pressure drop per meter of bed height	0.2–1.0 cm water gauge (max. 10 cm)
Inlet pollutant and/or odor concentration	0.01–5 g m⁻³, 500–50,000 OU m⁻³
Operating temperature	15–30°C
Inlet air relative humidity	>98%
Water content of the support material	60% by mass
pH of the support material	pH 6–8
Typical removal efficiencies	60–100%

Source: Deshusses, M.A., Biodegradation of Mixtures of Ketone Vapours in Biofilters for the Treatment of Waste Air, Ph.D. thesis, Swiss Federal Institute of Technology, Zurich, 1994.

are maintained. The system is properly aerated to ensure maximum degradation.

The free water phase benefits both biotrickling filters and bioscrubbers by providing a continuous supply of nutrients, removing possible toxic degradation by-products, suspending biomass for continual reseeding of the system, and aiding in the diffusion of hydrophilic pollutants into the biofilm.

1.2.7 Technology effectiveness and costs

There is no waste gas treatment technology that can effectively and economically be applied to every industrial or commercial application (Table 1.5). The effectiveness of a technology can often be defined by the flow rates and concentrations at which adequate cost-effective treatment can be expected (Figures 1.4 to 1.6). For all technologies, cost-effectiveness is site specific. Costs depend on the particular application, waste stream to be treated, materials required for construction, monitoring systems, etc. This makes it difficult to compare specific costs of technologies between sites. However, some general observations can be made.

Costs for waste gas control technologies vary because of processing differences. Because of energy costs, incinerators are best applied to air with higher concentrations of organics. Incinerators require large fuel inputs for low-concentration waste gases in order to ensure effective treatment. Incineration may also produce harmful by-products, such as NO_x, which contribute to other environmental problems (smog, acid rain, etc.). However, incinerators are insensitive to fluctuations, downtime, and the type of pollutant treated. A carbon adsorption system has high capital and operating costs

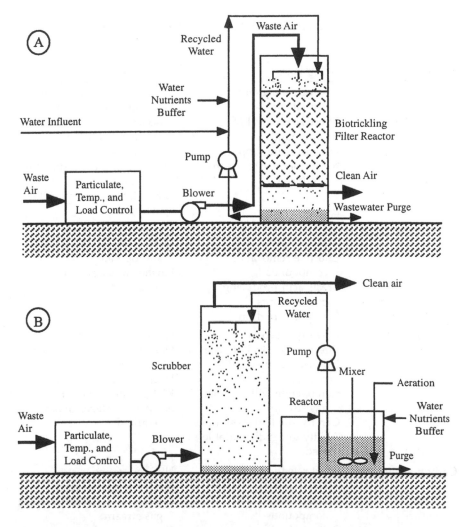

Figure 1.3 **(A)** Schematic diagram of biological trickling filter. Because the packing material is covered by a biofilm, absorption and biodegradation are achieved in the same reactor. The continuous reaction medium is the gaseous phase. **(B)** Schematic diagram of biowasher or bioscrubber. The absorption and biodegradation steps are separated. Settling and recycling of biomass are not shown in the diagram.

because of the expense of the medium and is appropriate for lower organic loading and solvents that are recoverable; however, this technology transfers the waste from an air to a solid phase that requires further treatment. On-site or off-site regeneration or reactivation of the spent carbon requires specially trained personnel and additional equipment. This will incur additional costs. Wet scrubbing may be suitable for some applications where odor control is necessary or if very corrosive gases are emitted. It requires

Table 1.5   Comparison of Waste Gas Control Technologies

Control technology	Advantages	Disadvantages
Biofiltration	Low operating and capital costs Effective removal of compounds Low pressure drop No further waste streams produced	Large footprint requirement Medium deterioration will occur Less suitable for high concentrations Moisture and pH difficult to control Particulate matter may clog medium
Biotrickling filters	Medium operating and capital costs Effective removal of compounds Treats acid-producing contaminants Low pressure drop	Clogging by biomass More complex to construct and operate Further waste streams produced
Wet scrubbing	Low capital costs Effective removal of odors No medium disposal required Can operate with a moist gas stream Can handle high flow rates Ability to handle variable loads	High operating costs Need for complex chemical feed systems Does not remove all VOCs Water softening often required Nozzle maintenance often required
Carbon adsorption	Short retention time/small unit Effective removal of compounds Suitable for low/moderate loads Consistent, reliable operation	High operating costs Moderate capital costs Carbon life reduced by moist gas stream Creates secondary waste streams
Incineration	System is simple Effective removal of compounds Suitable for very high loads Performance is uniform and reliable Small area required	High operating and capital costs High flow/low concentrations not cost effective Creates a secondary waste stream Scrutinized by public

Source: Webster, T.S., Control of Air Emissions from Publicly Owned Treatment Works Using Biological Filtration, Ph.D. thesis, The University of Southern California, Los Angeles, 1996.

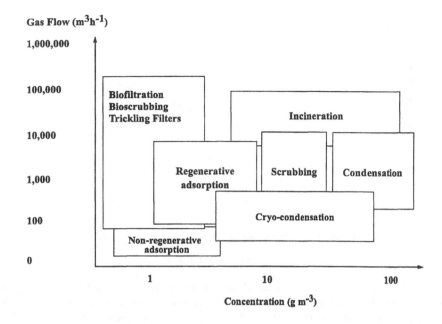

Figure 1.4 Applicability of various air pollution control technologies based on air flow rates and concentrations to be treated. (From Kosteltz, A.M. et al., in *Proceedings of the 89th Annual Meeting and Exhibition of the Air and Waste Management Association*, Pittsburgh, PA, 1996. With permission of KPMG Management Consultants, Ottawa, Ontario.)

no solid media disposal, but does generate wastewater and incurs high chemical costs. Membrane separation processes are capable of handling large VOC loads but are still partly experimental. A major drawback of the technology is the potential large electricity demand required to maintain significant pressure differential across the membrane in order to achieve high removal efficiencies. Biofiltration technology has low operating costs and produces minimal secondary pollutant waste streams; however, it may be inappropriate where high concentrations of organics or poorly degradable compounds are present. In general, biofilter technology is most cost effective for waste gas flows of 1000 to 50,000 m^3 h^{-1} and pollutant concentrations up to 1 g m^{-3} (Dragt, 1992; Kok, 1992).

1.3 Historical review of biofiltration

Microbial reactions have been used extensively to treat wastewater and solid waste throughout the twentieth century. However, it has only been since the 1950s that such techniques have been used to treat waste gases. Some of the earliest known biofilter systems were constructed as open pits filled with porous soil (Pomeroy, 1957). These pits had a simple air-distribution system

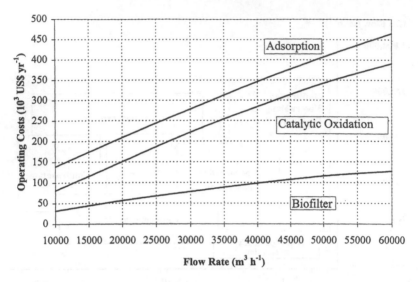

Figure 1.5 Investment costs vs. air flow rate for various air pollution control technologies. Costs are estimated (±20%) for the year 1997. Assumes 1 DM = $0.58 US. (Reprinted from Menig, H. et al., in *Biological Waste Gas Cleaning, Proceedings of an International Symposium*, Prins, W.L. and van Ham, J., Eds., VDI-Verlag GmbH, Dusseldorf, 1997. With permission.)

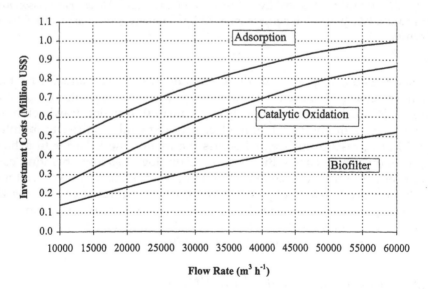

Figure 1.6 Operating costs vs. air flow rate for various air pollution control technologies. Costs are estimated (±20%) for the year 1997 assuming a 6.5% interest rate over 10 years. Assumes 6000 hours of operation per year and 1 DM = $0.58 US. (Reprinted from Menig, H. et al., in *Biological Waste Gas Cleaning, Proceedings of an International Symposium*, Prins, W.L. and van Ham, J., Eds., VDI-Verlag GmbH, Dusseldorf, 1997. With permission.)

of perforated pipes at the base of the soil, and the air was cleaned as it passed upward through the filter bed. Soil beds were often used to control sewer odor emissions generated at wastewater treatment plants (Pomeroy, 1957; Carlson and Leiser, 1966). These systems were generally successful, but the air distribution piping tended to clog and would deteriorate as acids were generated. Additionally, off-gases were unevenly distributed because of channeling caused by bed drying.

In the 1970s, interest in biofilters increased as stricter air quality regulations were enacted and enforced. More advanced biofilters capable of handling larger loads of odors and volatile organic compounds were needed. Primarily in Germany and the Netherlands, new systems were developed. These biofilters were open and had redesigned air-distribution systems. To avoid some problems of earlier biofilters, structural support media (bark, wood chips, polystyrene balls, etc.) were added to reduce filter bed medium compaction and to distribute the flow evenly. Though these changes improved biofilter performance, dry-out, compaction, and some acidification of the media were still observed.

During the 1980s and 1990s, biofiltration progressed rapidly in Europe and slowly in North America. Computer-operated, enclosed systems have been designed to treat odors, volatile organic compounds, and mixtures. Inorganic filter bed media have been extensively tested (Liu et al., 1994; Medina et al., 1995a,b; van Groenestijn et al., 1995; Graham, 1996). Granular activated carbon; powder activated, carbon-coated polystyrene; and ceramics have been used. Mixed with organic materials, they improve porosity while increasing the life expectancy of the filter bed. Additionally, the development of advanced design techniques and a mathematical model initiated by Ottengraf (1986) began the transformation of biofiltration research from a "black box" approach to a science-based effort.

As biofilter technology has been demonstrated and accepted by regulators as being a cost-effective, reliable means of controlling low-concentration biodegradable waste gases, new biofilter development and research companies have emerged in Europe and in North America. Some companies are developing novel biofilter designs, such as the Rotor-Biofilter (Sabo et al., 1996), to enhance biofilter performance. In addition, research is now being directed towards understanding pollutant biodegradation pathways, mixed pollutant treatment, transient behavior, nutrient limitation, inhibitors, biomass overgrowth suppression, and process modeling. It is only through such research that biofiltration can move forward to a more predictable engineering approach.

1.4 Biofiltration marketplace

The future of biofiltration technology is certainly contingent on the regulatory requirements placed on industry, but some general trends about the future viability of the technology can be discerned. The application of biofiltration technology throughout the world has increased rapidly during the latter part of the twentieth century and will continue to grow through the twenty-first century. Though recent studies vary based on assumptions made, the U.S. biofiltration market for the year 1996 was estimated to be about $10 million (Kosteltz et al., 1996). Projections for the U.S. market also vary, but economic models speculate that by the year 2000, the market may reach over $100 million (Yudelson, 1996). The industries which have shown the most promise for application of biofiltration technology include surface coating, municipal composting, pulp and paper, wood products, and site remediation (Kosteltz et al., 1996). In general, these industries fall within one of three potential markets: treatment of odors, treatment of VOCs and hazardous air pollutants (HAPs), or the treatment of petroleum hydrocarbons. In some instances, contaminant treatments may overlap.

Odor treatment is a significant portion of the marketplace. Industries that produce odorous emissions include composting facilities, foundries, fragrance manufacturers, bakeries, and wastewater treatment plants, among others. For example, composting facilities and wastewater treatment plants treat numerous forms of organic wastes, but the generation of odorous compounds causes local nuisance problems. Complaints from neighboring areas force these industries to implement some form of odor control. Odors such as sulfides, ammonia, mercaptans, alcohols, and ketones require control to limit the number of these complaints. Generally, composting and waste-water treatment plants produce large gas volumes of easily degradable compounds at low concentrations, making biofiltration an applicable technology (Paul and Sabo, 1995; Williams and Boyette, 1995). However, for processes such as composting, which are performed generally on a cost recovery basis, biofiltration technology may generate only small profits.

Biofiltration of VOCs and HAPs is gaining increasing popularity in the wood products, pulp and paper, and surface-coating industries. These industries generate a mixed-waste air stream. Odorous reduced sulfur compounds may be generated along with VOCs such as terpenes, methanol, phenol, formaldehyde, and chloroform. Surface-coating operations may also produce chlorinated compounds emitted from cleaning solvents. The trend for these industries is to replace raw process chemicals so that these odors and VOCs are not produced, but adequate substitutes are not always available. Gas volumes and concentrations will vary depending on the particular industry, but biofiltration shows great promise to treat these mixed streams of contaminants (Austin et al., 1995; Dombroski et al., 1995; Wolstenholme and Finger, 1995; Togna et al., 1997; Yavorsky, 1997).

Leaking underground storage tanks have created thousands of potential sites for the cleanup of soil and groundwater contaminated with gasoline, jet fuel, and diesel fuel. Biofiltration holds promise to treat air streams from soil vapor extraction or air sparging processes (Lesley and Chakravarthi, 1997).

Such air streams consist of hundreds of aromatic and aliphatic compounds. The aromatic portion — composed primarily of benzene, toluene, ethyl-benzene, and xylene (BTEX) — has been shown to be treated effectively using biofiltration (Chang and Yoon, 1995; Leson and Smith, 1995; Li, 1995; Stewart and Kamarthi, 1997; Wright et al., 1997). At these remediation sites, biofiltration may be combined with other air pollution control technologies to provide a rapid and cost-effective cleanup. Initially, these remediation sites generally have low gas flows with high contaminant concentrations. Incineration may be cost-effective for such contaminant streams; however, as concentrations decrease, the cost-effectiveness of incineration will decrease. Biofiltration can be cost effective in treating the low-flow, low-concentration waste streams. Biofiltration technology is often overlooked in this market because cleanup contracts are of short duration and require high removal efficiencies. As the mechanisms that control biofiltration effectiveness are becoming better understood, systems are being currently developed with rapid acclimation times and higher removal efficiencies.

1.5 Biofilter terminology

To describe the mechanisms of biofiltration clearly, general terminology pertinent to the field should be well defined. Because the field of biofiltration involves chemistry, microbiology, physics, fluid dynamics, and mathematics, much of the terminology found in this book has been taken from these fields. However, there are certain terms in the field of biofiltration that deserve special mention because of their common and sometimes confusing use within biofilter publications and presentations. Additional terminology pertinent to the field can be found in the glossary.

1.5.1 Empty bed residence time and true residence time

The term "empty bed residence time" (also "empty bed contact time" or "empty bed detention time") relates the flow rate to the size of the biofilter. It is defined as the empty bed filter volume divided by the air flow rate:

$$\text{EBRT} = \frac{V_f}{Q} \qquad (1.1)$$

where EBRT = empty bed residence time (seconds, minutes); V_f = filter bed volume (m^3, ft^3, etc.); and Q = air flow rate (m^3 h^{-1}, scfm, etc.)

The empty bed residence time overestimates the actual treatment time. The medium occupies a substantial fraction of the biofilter, reducing the volume within which the air flows and shortening the contact time. Even so, it is a commonly used parameter because it is easily calculated. The true residence time, which is the actual time a parcel of air will remain in the biofilter, is defined as the total filter bed volume multiplied by the bed porosity of the filter medium, divided by the air flow rate:

$$\tau = \frac{V_f \times \theta}{Q} \tag{1.2}$$

where τ = true residence time (seconds, minutes); θ = porosity = volume of void space / volume of filter material.

In literature, the terms "empty bed residence time" (EBRT) and "true residence time" (τ) are both commonly used. The difference between these two terms is the porosity factor (θ) and can be quite substantial. Hence, care must be taken by the reader to understand which definition applies to a particular situation.

The effects of the empty bed residence time or true residence time on the performance of a biofilter are parallel. Generally, as either EBRT or τ increases, either by reducing the volumetric flow rate or by increasing the volume of medium, the system performance will improve. For many particular biofilter sites, the flow rate is fixed and a function of the contaminant producing process. Hence, reactor volume is often the only variable that can be increased. Increasing the porosity or size of the filter bed can increase this volume. However, biofilters with larger volumes and longer gas residence times are more expensive. Typical vapor residence times for commercial and industrial applications range from 25 seconds for the treatment of odor and low VOC concentrations to over a minute for high concentrations of VOCs (Leson and Winer, 1991).

1.5.2   Surface (or volumetric) and mass loading rate

Surface (or volumetric) and mass loading rate are terms used to define the amount of air or contaminant that is being treated. Both terms are normalized, allowing for comparison between reactors of different sizes. Surface loading rate is defined as the volume of gas per unit area of filter material per unit time (in metric units as m^3 of gas per m^2 of bed surface per hour). Similarly, the volumetric loading rate is defined as the volume of gas per unit volume of filter material per unit time (in metric units as m^3 of gas per m^3 of filter material per hour).

$$\text{Surface loading} = \frac{Q}{A} \tag{1.3}$$

where A = filter area (m^2, ft^2).

$$\text{Volumetric loading} = \frac{Q}{V_f} \tag{1.4}$$

The mass loading rate (either surface of volumetric) is the mass of the contaminant entering the biofilter per unit area or volume of filter material per unit time, often expressed as grams per m² or m³ of filter material per hour. Because flow remains constant through a filter bed, the mass loading along the length of the bed will decline as contaminant is removed. However, generally an overall mass loading rate for a system is defined:

$$\text{Mass loading (surface)} = \frac{Q \times C_{Gi}}{A} \qquad (1.5)$$

where C_{Gi} = inlet concentration (g m^{-3}).

$$\text{Mass loading (volumetric)} = \frac{Q \times C_{Gi}}{V_f} \qquad (1.6)$$

1.5.3 Removal efficiency and elimination capacity

Removal efficiency and elimination capacity are used to describe the performance of a biofilter. Removal efficiency (RE) is the fraction of the contaminant removed by the biofilter, expressed as a percentage:

$$\text{Removal efficiency} = \left(\frac{C_{Gi} - C_{Go}}{C_{Gi}}\right) \times 100 \qquad (1.7)$$

where C_{Gi} = inlet concentration (ppmv, g m^{-3}); C_{Go} = outlet concentration (ppmv, g m^{-3}).

Elimination capacity (EC) is the mass of contaminant degraded per unit volume of filter material per unit time. Typical units for elimination capacity are grams of pollutant per m³ of filter material per hour. An overall elimination capacity is generally defined:

$$\text{Elimination capacity} = \frac{(C_{Gi} - C_{Go}) \times Q}{V_f} \qquad (1.8)$$

$$\text{Elimination capacity} = \text{Volumetric mass loading} \times \text{RE} \qquad (1.9)$$

Removal efficiency is an incomplete descriptor of biofilter performance because it varies with contaminant concentration, airflow, and biofilter size and only reflects the specific conditions under which it is measured. The elimination capacity allows for direct comparison of the results of two different biofilter systems because the volume and flow are normalized by definition; however, elimination capacity is also a function of input concentrations.

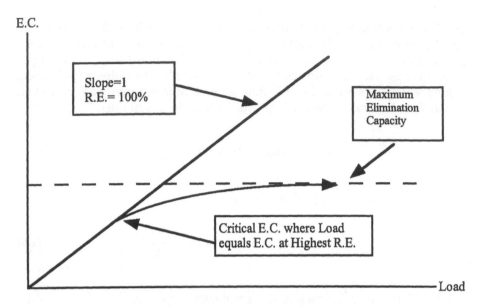

Figure 1.7 Typical elimination capacity vs. load curve. Elimination capacity is always equal to or less than the load. The ratio between elimination capacity and load is the removal efficiency of the system.

Effluent concentration (or percent removed) is still commonly used as the goal of regulatory compliance.

Elimination capacity can only be equal to or less than the mass loading rate. Under low load conditions, the elimination capacity essentially equals the load, and the system is calculated to be at 100% removal efficiency (Figure 1.7). By increasing the load on a system, a point will be reached where the overall mass loading rate will exceed the overall elimination capacity, generating removal efficiencies less than 100%. This point is typically called the critical load or critical elimination capacity. The decline in removal efficiency may be explained differently depending on which parameter is increased to increase the overall mass loading rate. If the flow rate is increased or the volume decreased, the residence time is reduced, and the contaminant may not have sufficient time to diffuse into the biofilm and be readily oxidized. Conversely, if the concentration is increased and the flow rate and volume remain the same, the biofilm may not be able to absorb the increase in concentration, with some of the contaminant simply passing through the system untreated. As the loading rate continues to increase, a maximum overall elimination capacity (EC_{max}) will eventually be reached. This maximum overall elimination capacity is independent of contaminant concentration and residence time within a reasonable range of operating conditions. Elimination capacities for conventional biofilters treating common pollutants typically range from 10 to 300 g m^{-3} h^{-1}. Published elimination capacities are shown in Appendix B. In general, a meaningful description of intrinsic biofilter performance should include C_{Gi}, Q, V_f, RE, and EC.

1.6 Additional biofiltration resources

This book is by no means the only source of information on biofiltration. In recent years, a few chapters in environmental engineering books or encyclopedias have been written, primarily in Germany, the Netherlands, and the U.S., which also discuss various aspects of the field. Noteworthy contributions include "Exhaust Gas Purification" in *Biotechnology* (Ottengraf, 1986), VDI-Berichte Guideline 3477: "Biological Waste Gas/Waste Air Purification, Biofilters" (VDI-Berichte, 1991), "Process Technology of Biotechniques" in *Biotechniques for Air Pollution Abatement and Odour Control Policies* (Ottengraf and Diks, 1992), and "Biofilters" in the *Handbook of Odor and VOC Control* (Skladany et al., 1998). Portions of Chapters 2, 4, 5, and 7 of this book have appeared previously in the article, "Biofiltration Technology for Air Pollution Control", published in the *Encyclopedia of Environmental Analysis and Remediation* (Webster and Devinny, 1998). The *Journal of Environmental Engineering*, published by the American Society of Civil Engineers, has devoted an issue entirely to the biofiltration of air (Vol. 123, No. 6, 1997). These resources offer insight into the developing field of biofiltration and detail some of the principles and applicability of biofiltration for waste air treatment. However, not all of these resources are readily available in the U.S., and access is often difficult.

Because of the rapid maturation of the field, most resources on biofiltration quickly become outdated. For this reason, the most current information on biofiltration research and development must be obtained through yearly conference proceedings or monthly scientific periodicals. For conference proceedings, there are generally four to five yearly or bi-yearly conferences which offer the most timely information. The annual conferences include the Air and Waste Management Association Conference and the conferences held by the American Institute of Chemical Engineers, the American Chemical Society, and the American Society of Civil Engineers. Bi-annual conferences include the VDI/LUCHT Biological Waste Gas Cleaning Symposium, the University of Southern California/The Reynold's Group (USC/TRG) Conference on Biofiltration, and the Battelle *In Situ* and On-Site Bioremediation Symposium. The proceedings from such conferences offer a wide array of information pertaining to the general development of the technology and more complex research.

Numerous scientific journals, which regularly present biofiltration articles, may also be a source for current biofiltration trends and developments. Journals that publish work on biofiltration include *Biotechnology and Bioengineering, Environmental Science and Technology, Environmental Progress*, the *Journal of the Air and Waste Management Association*, and the *Journal of Environmental Engineering*. Many other journals present articles detailing the progress of the technology, but the previously mentioned journals are a suitable base of information from which other resources may be found.

Finally, the use of the Internet as a source for rapid information dissemination has become a very powerful tool in the field of biofiltration. Biofilter information exists on the Internet in terms of paper abstracts and commercial literature. Most of this biofilter literature can be found through environmental journal and symposium web pages. The World Wide Web page addresses for such information can be found in their respective journals or through a general information search.

1.7 Conclusions

Through the twenty-first century, increased production by the commercial and industrial sectors will lead to further waste gas control regulations. Today, the strength of the standards only appears to be increasing as more harmful effects of air pollution are discovered. These trends are forcing companies to look for methods to control waste gas production while maintaining profits. Product substitution and recycling are possible for some companies, but many others will find increasing need for economical, effective waste gas treatment technologies. Traditional technologies such as incineration and absorption are effective air pollution control technologies. However, the economic effectiveness of incineration declines when it is used for low-concentration waste streams, while carbon adsorption often creates secondary wastes that must be treated as solid hazardous waste. An alternative to these traditional technologies has been the development of biofiltration technology. Since the 1970s, the field of biofiltration has been growing and becoming more prominent in the waste gas control market because of its simplicity, effectiveness on dilute waste streams, and low costs. Through its short history, the technology has advanced from treating odors with simple open soil beds to treating volatile organic compounds using enclosed, computer-controlled, multi-bed systems. Advancements in the technology have focused on better control of operational conditions and a detailed understanding of kinetics and removal mechanisms. As the technology advances, further research in the field will transform it from an empirical practice to one more deeply rooted in concrete theory and scientific principles.

chapter two

Mechanisms of biofiltration

2.1 Introduction

The underlying mechanisms which allow biofilters to work, and which must be controlled to ensure success, are complex. The biofilter contains a porous medium whose surface is covered with water and microorganisms. Treatment begins with transfer of the contaminant from the air stream to the water phase. The dissolved contaminant is moved by diffusion and by advection in the air (biofilters are presumed to see little water flow in comparison to biotrickling filters, but some percolation may occur). The contaminant may form complexes with organic compounds in the water. It may adsorb to the exopolysaccharides released by the biofilm-forming cells or to the cells themselves. It may also be adsorbed by the support medium. Ultimately, biotransformation converts the contaminant to biomass, metabolic by-products, or carbon dioxide and water. If the contaminant contains chlorine or sulfur, these will appear as chloride and sulfate. The biodegradation is carried out by a complex ecosystem of degraders, competitors, and predators that are at least partially organized into a biofilm.

Each of the steps in the biofiltration process should be understood. Each can be interrupted, causing biofilter failure, and each provides opportunities for improvement in biofilter operation. Because the steps are largely sequential, it is often important to identify the slowest step, which limits biofilter efficiency.

2.2 Gas transfer

2.2.1 The equilibrium

Movement of the contaminant from the air to the water phase occurs according to physical laws which are familiar to environmental engineers. At equilibrium, the partition between the air and water is described by Henry's

Law. Concentrations in the water will be proportional to those in the air, and the constant of proportionality is the Henry's constant:

$$C_G = HC_L \tag{2.1}$$

where C_G = the concentration of contaminant in the air phase, atm, or g L_{air}^{-1}; C_L = the equilibrium concentration of contaminant in the water phase, mol L_{water}^{-1} or g L_{water}^{-1}; H = the Henry's Law constant, atm L mol^{-1}, or g L_{air}^{-1} per g L_{water}^{-1} (dimensionless).

A key concern for understanding implications of Henry's Law for biofilters lies in the units. Constants are typically given in atmospheres per mole per liter, relating the gas partial pressure to the water concentration. These units are convenient for many applications and generally produce numbers which are close enough to one that they can be expressed without scientific notation. However, these units tend to hide an important point: concentrations in the air are commonly much lower than those in the water. Expressed in units of g L^{-1} of air over g L^{-1} of water (dimensionless), Henry's constants are almost all well below 1, even for compounds we recognize as hydrophobic. Constants for hydrophilic compounds are very low.

This means that for any volume within a biofilter, more contaminant is likely to be in the water than in the air. This contributes to retardation of the contaminant as the air moves through the biofilter and ultimately makes biofilters workable. The air typically remains in the biofilter for only a minute or so. If the contaminant were not retained for much greater times, biodegradation could not occur. This partition and retardation effect is compounded by the adsorption of contaminant by biomass and the support medium. Retardation is determined by the total partition coefficient K_{mass}, that is, the ratio between the mass of material in the air and the mass in all the other phases, measured within a given volume of biofilter. For many combinations of compound and support medium, the mass adsorbed will be much greater than the mass dissolved in the water, and these two effects together can produce high retardation factors.

Equilibrium is a local phenomenon. In biofilters, concentrations of contaminant decline substantially as the air moves from the regions near the inlet to regions near the outlet and will be lower deep in the biofilm than they are near the air-water interface. It is possible that the concentrations in the air and water will be near equilibrium throughout the biofilter, even as both concentrations differ strongly from place to place.

2.2.2 *Transfer rates*

Biofilters are economical only if the air can be passed through them quickly. Transfer of the contaminant to the medium must also be fast. Pollutant mass transfer from air to water is of prime importance in many environmental

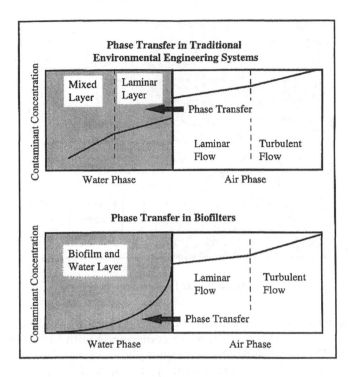

Figure 2.1 Models for gas transfer.

applications, and has commonly been modeled in four steps (Figure 2.1). Consumption of a compound in the water phase produces low concentrations, so that transfer occurs from regions of high concentration in the air to regions of low concentration in the water. The bulk of the air flow is turbulent, so that the contaminant moves by advection and eddy diffusion. The combination is usually referred to as "convection". Near the air-water interface, the air flow becomes laminar, so that molecular diffusion becomes the only transport mechanism. Because molecular diffusion is slower than convection, this can be the rate-limiting factor for the mass transfer into activated sludge or other aqueous systems, especially in the case of transfer of highly soluble compounds (gas-phase control condition). But air flow is rapid in biofilters, so the laminar layer is kept thin. Because other processes are slow, transfer in this layer is not limiting in most cases. However, the shape of the water surface in biofilters is highly irregular. It is likely that transfer hindrance occurs in laminar air within partially isolated pockets and corners throughout the biofilter (Figure 2.2). The magnitude of this effect has not been measured.

In traditional systems requiring gas-liquid mass transfer, the laminar water layer in the aqueous phase may also be a limiting factor, slowing

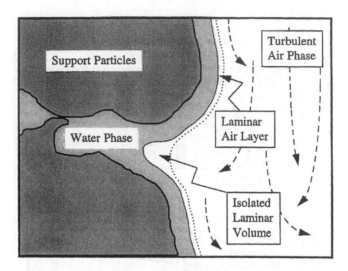

Figure 2.2   Air flow in biofilters.

transfer of either the contaminant or of oxygen to the bulk water phase (Figure 2.1). This is especially true for compounds with high Henry's Law coefficients. This condition is called "liquid phase control" because diffusion through the water film is the rate-limiting step. In biofilters, there is no bulk turbulent phase because the water is essentially stationary. Indeed, most of the water may exist as a water-saturated biofilm. Diffusion through the biofilm may be further hindered by cells and exopolysaccharides. In many cases, the biofilm occupies the entire water volume, so there is no free water layer.

Interphase mass transfer (from the air to the water) is presumed to occur at a rate that is proportional to the degree to which the concentration in the water is below the equilibrium value. According to the two-film theory, the rate of interphase mass transfer can be calculated using bulk concentration in the liquid (or gas) phase and overall mass transfer coefficients:

$$\frac{dC_L}{dt} = k_t(C_L^* - C_L) = k_t\left(\frac{C_G}{H} - C_L\right) \tag{2.2}$$

where C_L = the concentration of contaminant in the bulk water; C_L^* = the contaminant concentration at equilibrium with local air concentration; t = time; k_t = a transfer rate constant, per unit time, also equal to K_La, where a is the interfacial area and K_L the overall mass transfer coefficient based on overall driving force in the liquid.

The constant in the equation depends on several factors. Different compounds will have different rate constants, but a fundamental factor is the interfacial surface-to-volume ratio. Transfer occurs through the surface, and, with other factors held constant, the amount of mass transferred will be proportional to the amount of surface available.

The mass of contaminant which is lost from the air is the same as the mass which is gained by the water, but the effects on the concentrations will be different, because the volumes of the air and water phases are different:

$$\frac{dC_G}{dt} = -k_t \frac{V_L}{V_G}(C_L^* - C_L) \qquad (2.3)$$

where V_L = the volume of the water phase; V_G = the volume of the air phase.

This model is highly simplified because it presumes that mass transfer rates are uniform through the various air and water interfaces (k_t assumed constant). In reality, there may be significant differences in the specific resistances to mass transfer depending on the position in the biofilter, because both the air flow pattern and biofilm geometry are changing. For example, the water phase in a biofilter is highly non-uniform. A thin layer adsorbed to the surface of a convex particle may reach equilibrium very quickly, while a neighboring volume of water in a pore is substantially out of equilibrium. Biological activity is constantly depleting aqueous concentrations of contaminant, and it may not do so uniformly. The cells and exuded organic materials of the biofilm may completely occupy the water phase, substantially slowing diffusion away from the interface.

Even so, the simple two-film linear model is widely used. In many instances, it is further simplified to reflect the fact that resistance to gas-phase diffusion is often negligible compared to liquid-phase diffusion (liquid-phase control situation). In such cases, it is assumed that concentrations at the surface of the water phase can be assumed to be at equilibrium with the air (one-film theory). Using this assumption and calculating concentrations within the water based on other principles may effectively deal with the complexity of gas transfer, allowing biological reactions to be readily included and the differential equations for mass transfer to be solved either numerically or analytically.

Determination of transfer rate coefficients is difficult because measured transfer rates are also affected by biodegradation rates and the mass partition coefficients. However, the mass partition coefficient can be measured readily in a pulse test (described in the following section). The effects of biodegradation can be minimized for an experiment by adding a biological inhibitor to the biofilter and operating it at very high surface loads. Under these conditions, measurements of the output pollutant concentrations during the breakthrough period can be fitted to a simple model to estimate the transfer rate coefficient (Hodge and Devinny, 1997).

2.3   The water phase

In biofilters, in contrast with biotrickling filters, it is assumed that the water phase is stationary. Downward movement may occur if condensation or irrigation is heavy, but the flow is slow and operators generally minimize it in order to avoid producing leachate. Water movement is always laminar rather than turbulent. Under these conditions, diffusion will be the dominant means by which contaminant moves from place to place in the water. Concentrations are higher near the surface of the water layer, where transfer from the atmosphere is occurring. Biodegradation in the water or biofilm and adsorption at the surface of the support medium act as sinks. Contaminant diffuses towards the support, and the products of biodegradation diffuse outwards. Those products with high vapor pressures are transferred into the air.

Diffusion in water or biofilm is commonly orders of magnitude slower than in air, and transfer resistance within the water phase is likely. Contaminant concentrations may be depleted at the bottom of the water layer. Oxygen may be absent, and metabolic intermediates may accumulate. Design and modeling of biofilters must take into account the fact that only the surface portion of the biofilm may be active.

2.4   Adsorbed contaminants

The transfer of contaminants from the air to the water and solids in a biofilter is a fundamental step in treatment and is sometimes loosely referred to as adsorption or dissolution. However, it ultimately involves several complex mechanisms. Contaminant molecules may be simply dissolved in the water, but they may also be adsorbed on the surface of the medium, taken up by living cells, adsorbed on the surface of biofilm organic matter, absorbed within organic matter in the biofilm or medium, or collected at the surface of the water (Figure 2.3). For highly soluble contaminants such as ethanol, the dissolved form may be dominant, and the volume of the water phase will have considerable influence on the amount transferred from the air (Hodge and Devinny, 1997). For more hydrophobic contaminants, the major reservoir may be material adsorbed on the surface of the medium and absorbed within the organic matter.

Adsorption phenomena in biofilters are poorly understood but important to biofilter operation. The ratio of the total amount of contaminant in the water-and-solids phase to the amount in the air determines the residence time of the contaminant in the biofilter. Contaminants at the surface of the medium or in large pores may be available for biodegradation, while those in pores too small for microorganisms may not be. Contaminants vary in their affinities for water, medium, and organic matter, and a given contaminant will adsorb differently on different media. Adsorption and desorption

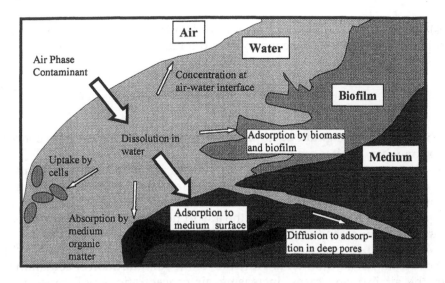

Figure 2.3   Adsorption in biofilters.

from different media or from different parts of a single medium will occur at different rates. The general goal of biofilter design in this respect is to achieve the maximum possible concentration of contaminant in the forms that are available for biodegradation. The best way to do this is not always obvious.

Adsorption on solid media from water has been widely studied in the fields of water treatment and soil bioremediation. Two simple models of adsorption at equilibrium are commonly used. In the Freundlich model, it is assumed that sites for adsorption are not limited, and the amount of contaminant held will depend on the concentration in the water:

$$C_{ads} = k_f C_L^{1/n} \qquad (2.4)$$

where C_{ads} = concentration of adsorbed contaminant; C_L = concentration in the liquid phase; n = a constant; k_f = Freundlich adsorption constant.

In its most basic form, the model is linear ($n = 1$), but nonlinearity is observed in some cases and can be accommodated in the model by varying n (but numbers near 1 are typical). The Freundlich model implicitly assumes that the adsorption capacity is unlimited in the range over which the relationship is applied: increasing the liquid phase concentration will always increase the amount of contaminant adsorbed.

Other models assume that adsorption occurs at specific sites and that each site can be occupied by only one contaminant molecule. The Langmuir relationship can be calculated assuming chemical equilibrium between dissolved and adsorbed concentrations, and a limited number of sites:

$$C_{ads} = \frac{C_{max}C_L}{k_L + C_L}$$ (2.5)

where k_L = the Langmuir adsorption constant; C_{max} = the maximum concentration when all sites are occupied.

If C_L is small, most of the adsorption sites are not occupied, adsorption is not limited by the total number of sites, and the Langmuir relationship approximates the linear Freundlich form. When the concentration in the liquid is very high, essentially all the adsorption sites will be occupied, and the amount adsorbed will be a constant independent of concentration.

The biofilter operator must be aware that the "adsorption capacity" of the medium in the biofilter is a function of the concentration of contaminant in the air in many cases. Increasing air concentrations will cause more contaminant to adsorb, and a drop to lower concentration will cause it to be released. The biofilter medium may serve as a buffer for concentration changes.

Adsorption and desorption do not occur instantaneously. In some cases they may be very fast, so that systems are always near equilibrium, and the Freundlich or Langmuir relationships are always satisfied. For other combinations of contaminant and adsorbent, adsorption and desorption may be very slow. It has long been recognized, for example, that degradation rates of contaminants adsorbed on soils may be limited by desorption rates, and that these may be low enough to require years for completion of remediation projects. Slow desorption may occur, for example, when contaminants are held in deep pores with narrow openings, so that diffusion out to the surface of the particle is slow.

When adsorption rates are slow, the amount of contaminant held may be different from the adsorption capacity, which is the amount which would be held at equilibrium. Because adsorption capacity varies with concentration, a rapid change in contaminant concentration can leave the medium holding amounts above or below the equilibrium adsorption capacity, and desorption or adsorption will occur. Under these conditions, contaminant in the water phase will be changing as a result of both biodegradation and sorption, and both must be included in efforts to model biofiltration dynamics.

The effects of adsorption on biodegradation rates are complex, depending on the medium, contaminant, and microorganisms (Alexander, 1994). Sometimes the adsorbed contaminant is simply unavailable to the microorganisms. It may be adsorbed in pores too small for microorganisms to enter, or it may be firmly bound in a way that prevents uptake. Under these conditions, biodegradation at steady state will not be affected. The mass of contaminant that was initially adsorbed will remain, and longer term operation

Table 2.1 Summary of Effects of Adsorption on Biodegradation

Compound	Rates	Results
Unavailable; in small pores, active site for enzyme blocked	Fast desorption	Adsorbed material will not affect steady-state rates, but will buffer changes in input concentration, helping to stabilize operation
	Slow desorption	Only small effects on operation over long term, which may not be noticeable; may make disposal of medium more difficult
Partially available; pore sizes vary; multiple adsorption mechanisms	Fast desorption or slow desorption	The available portion will contribute as described below; the unavailable portion will contribute as described above
Available; active site available to enzymes; microorganisms have surfactants or exoenzymes	Desorption rate may not be significant	Available adsorbed material will raise concentration available for degradation and raise degradation rates

will occur just as if the adsorbed material did not exist. However, if desorption is rapid and input concentrations vary, adsorption of additional material while concentrations are rising may reduce toxic shock. Desorption while concentrations are falling will make substrate available, keeping the microorganisms healthy and degradation rates high (Table 2.1). Because varying input concentrations are the rule rather than the exception, this buffering effect is commonly valuable.

If adsorption and desorption rates are slow, an unavailable reservoir of adsorbed contaminants will have little effect. Over long periods of operation, contaminant will gradually accumulate, and if operation is interrupted, contaminant will be gradually released. Typically, media are decontaminated for disposal by running the biofilter on air only until all of the adsorbed contaminant is degraded. But, an unavailable reservoir of hazardous contaminant could conceivably cause the medium to be declared a hazardous waste, making disposal expensive.

The adsorbed contaminant may be available for biodegradation. This can occur if exoenzymes are released by the microorganisms and can degrade adsorbed material to produce desorbed products. The exoenzyme might be a stereospecific protein that attaches to the substrate at a specific site, which succeeds because the site is on an exposed portion of the adsorbed molecule. Other exoenzymes are non-specific. Some degrade organic compounds by releasing peroxide. Presumably these can oxidize compounds while they are adsorbed, just as they are used by cells to break down natural substrate such as wood particles that cannot be taken into the cell. In some biofilters, investigators have used white rot fungi, noted for the effectiveness of their exoenzymes (Braun-Lullman et al., 1995). Molecules adsorbed to cells or biopolymers will likely be available. Many microorganisms release surfactants that can promote desorption of contaminants.

Exoenzymes or biosurfactants will be present only if the species of microorganisms that produce them are present. Thus, the microbial ecology can influence the physical chemistry of the system, and this can change with time. Alexander (1994) reported that an initial culture of microorganisms could not degrade biphenyl adsorbed on polyvinylstyrene beads, but that inoculation with sediment and 14 days of acclimation allowed proliferation of a culture that could do so.

The adsorption and release of contaminant can have another substantial effect on biofilter operation. It gives the biofilter a "retardation coefficient". The velocity of contaminants in the biofilter is the weighted average of the velocities of the airborne and adsorbed masses:

$$v = \frac{v_G M_G + v_{ads} M_{ads}}{M_G + M_{ads}} \qquad (2.6)$$

where v = the average contaminant velocity; v_G = the interstitial velocity of the air; v_{ads} = the velocity of contaminant that is in the water or on the solid; M_G = the mass of contaminant per unit volume of biofilter in the air; M_{ads} = the mass of contaminant in the water or on the solid per unit volume of biofilter. But, the contaminant in the water or on the solid is not moving (v_{ads} = 0), so:

$$v = \frac{v_G M_G}{M_G + M_{ads}} = \frac{v_G}{R} \qquad (2.7)$$

where R, the ratio of total mass of pollutant per unit biofilter volume to the mass per unit volume in the air, is called the "retardation factor". The passage of the contaminant through the biofilter is slowed by adsorption, and the detention time of the contaminant will equal the air detention time multiplied by R. Values of R typically vary from two or three to tens of thousands (Hodge and Devinny, 1994).

The retardation factor is closely related to the mass partition coefficient:

$$R = \frac{M_G + M_{ads}}{M_G} = 1 + \frac{M_{ads}}{M_G} = 1 + K_{mass} \qquad (2.8)$$

where K_{mass} is the ratio between the mass of contaminant adsorbed and the mass of contaminant in the air at any point in the biofilter. If R is measured, K_{mass} is known, and for most cases R is almost equal to K_{mass}. Direct measurement of K_{mass} can be difficult. A known amount of contaminant can be placed in a closed vessel with a known amount of active biofilter material, and the concentration and amount of contaminant in the air can be measured. Biodegradation will be consuming the contaminant, however, so measured concentrations will be changing. This interference may be reduced by poisoning the biomass, but killing it completely is surprisingly difficult because biofilms on particulate material are quite rugged. Autoclaving is effective, but it may change the nature of the medium to such a degree that the adsorption data are no longer relevant to the real biofilter. Irradiation may be the best method, but the necessary equipment is not always available.

Measurement of R is much easier. In an active biofilter under real conditions, it is possible to induce a spike in the inlet concentrations. This produces a peak of concentration which moves through the biofilter just as a peak moves through a gas chromatograph. Its velocity can be determined by recording the time which passes before the peak appears at the outlet. The ratio of the interstitial velocity of the air to the speed of movement of the peak is the retardation factor (Hodge and Devinny, 1995). It is also necessary to be aware of the "peaks" and "valleys" of concentration passing through the biofilter, because they will complicate measurement of removal efficiency (see Chapter 8).

The adsorptive capacity of the dry raw material used in a biofilter will be substantially reduced as the biofilter is put into operation. Water will compete for adsorptive sites, because the medium is always kept wet (Loy et al., 1997). Biopolymers may also coat the surface, preventing the attachment of contaminant molecules. Mohseni and Allen (1996) found that the adsorptive capacity of carbon fell to values near those for perlite after a few months of biofilter operation. The true retardation factor reflects the combined effects of dissolution in the water, adsorption or absorption in the biomass, and adsorption to the medium.

New biofilter material often provides good treatment for the first few days of operation, because it is acting as an adsorber (Mohseni and Allen, 1996; Medina et al., 1995b). These excellent results will not continue once the bed adsorptive capacity is occupied, and removal depends on biodegradation only. However, when a new biofilter is being started, biodegradation of the contaminants may also be delayed by the need for the microbial ecosystem to become acclimated (see Chapters 5 and 8). Adsorption which occurs during this time may be helpful in that it reduces the amount of contaminant which escapes during the microbiological acclimation period (Bishop and Govind, 1995). A similar effect may occur when there is a sudden increase in the input concentration. Further, if the input concentration falls at night or on weekends when contaminant generation ceases, the adsorbed contaminant may be released, keeping the microorganisms active. Overall, the adsorptive capacity of the medium tends to smooth or buffer changes in concentrations and may reduce stress on the microbial population.

2.5 Contaminant biodegradation

2.5.1 The biofilm

The key element in destroying contaminants is the biofilm. This is a mass of organisms growing on the surface of the solid medium and carrying out the metabolic activities which transform the contaminant to harmless products. If the biofilter is to succeed, these organisms must be very active. A greater rate of transformation allows the biofilter to be smaller and less expensive, and necessary biofilter size often determines economic success.

The structure of biofilms in real reactors is poorly known and is the subject of continuing investigation (see Chapter 5). Viewed from a distance, the biomass is a thin film at the interface between the solid and the gas. It may be from a fraction of a millimeter thick to 5 mm thick in successful biofilters, and it grows until it clogs pores 2 cm in diameter in overloaded biofilters. At the microscopic scale, many structures are possible, and it is probable that different biofilters support much different biofilms. The biofilm may be a smooth, relatively uniform layer of cells embedded in a polysaccharide gel that is mostly water. But, there is evidence that it may also be penetrated by channels which allow water to flow among clumps of biomass. There may be free water outside of the biomass which allows swimming organisms to thrive. Filamentous species may protrude above the water into the air (Hugler et al., 1996; Møller et al., 1996; see also Chapter 5).

In any case, many of the microorganisms present are busily collecting contaminant molecules and metabolizing them. In most cases, they are doing so in order to utilize the chemical energy the contaminant contains. Some carbon from the contaminants may be used by the cells for their own growth, and some will be released again as they are consumed by predators. Most of the time, investigators refer to this complex microbial ecosystem simply as the biofilm, but an awareness of its true complexity should be retained.

2.5.2 Kinetics

The microbial degradation which occurs in biofilters is similar to that in bioremediation of soils or biological treatment of wastewater. The kinetics of contaminant degradation are often modeled using the Michaelis-Menten equation, developed for enzyme mediated reactions:

$$\frac{dC_L}{dt} = \frac{k_{max}C_L}{K_s + C_L} \tag{2.9}$$

where C_L = contaminant concentration in the liquid, mol L^{-1}; k_{max} = maximum degradation rate, mol L^{-1} s^{-1}; K_s = half-saturation constant, mol L^{-1}.

Where concentrations are high with respect to the half-saturation constant, the biodegradation rate is approximately equal to k_{max} and will not change with concentration (degradation will be zero-order). Where concentrations are well below the half-saturation constant, the biodegradation rate is approximately proportional to the contaminant concentration (degradation will be first-order). It is quite possible that the kinetics will be zero-order near the biofilter inlet, where concentrations are high, but first-order farther along in the biofilter where concentrations are low.

Application of the Michaelis-Menten relationship to whole cell systems assumes that the number of microorganisms present is not changing, so that k_{max} and K_s do not change. In many cases, however, the biomass will be developing as contaminant is consumed. Because contaminant degradation is the result of microbial activity, the kinetics of contaminant degradation are closely related to the kinetics of microorganism growth. Growth is commonly presumed to be proportional to the size of the microbial population:

$$\frac{dX}{dt} = \mu X \tag{2.10}$$

where X = the density or concentration of biomass, mg L^{-1}; μ = the coefficient of proportionality, specific growth rate, s^{-1}.

The coefficient of proportionality, or growth constant, in turn depends on the concentration of contaminant available for use by the microorganisms:

$$\mu = \frac{\mu_{max} \times C_L}{K_m + C_L} \tag{2.11}$$

This is the Monod relationship, describing the rate at which biomass is expected to grow. Thus, just as contaminant degradation is first-order at low concentration and zero-order at high concentration, the growth rate will rise linearly at low contaminant concentration and become constant at high contaminant concentration.

In some cases, it is known that very high concentrations of substrate can become inhibitory. One relationship used to model this situation is the Haldane equation:

$$\mu = \frac{\mu_{max} C_L}{K_S + C_L + \left(C_L^2 / K_i \right)} \tag{2.12}$$

where K_i = inhibition constant. The squared term in the denominator describes a substantial inhibitory effect which will occur as C_L rises to values near the square root of K_i.

Alexander (1994) has described these three kinetic equations along with others for biodegradation in soils. He has also noted several reasons why kinetics in real systems may depart from these rules. Most apply to biofilters as well as soils, and some can be summarized here:

1. Diffusional barriers may exist.
2. The substrate may be sorbed on the medium.
3. The degrading microorganisms may be helped or hindered by the presence of other compounds.
4. Inorganic nutrients or oxygen may be limiting, rather than the substrate.
5. Several species, with different degradation constants, may be active.
6. Predators may limit the numbers of bacteria.
7. Cells may be aggregated, affecting substrate transport processes.
8. Acclimation of the microbial population may be necessary.

The diffusional limitation suggested in (1) is likely a common factor in biofilters. Degradation kinetics are further complicated because at low concentrations the contaminant may not diffuse into the biofilm rapidly enough to penetrate the full depth of the biofilm. Under such conditions, less biomass will be utilized and less substrate consumed as the concentrations decline. Degradation rates will be limited by diffusion phenomena rather than by biological activity. Ottengraf's model (Ottengraf, 1986) predicts rates will be a quadratic function of contaminant concentration.

Aerobic processes dominate in biofilters because large amounts of air flow through them. Contaminants are generally present in concentrations low enough so that even their complete oxidation does not deplete the oxygen available in air. Aerobic degradation is desired because it is generally faster and less likely to produce objectionable by-products. Anaerobic conditions may occur at the bottom of biofilms or within remote pores in the support medium where the supply of oxygen is diffusion limited. This occurs if degradation rates are high and oxygen consumption within the biofilm exceeds the rate at which it can diffuse inward through the water layer.

In a poorly operated biofilter, where there is too much water, or the support medium is decomposing and compacting, the air flow may be channeled. Larger regions of the biofilter may become anaerobic because the air flows around them. Anaerobic conditions may generate sulfides, mercaptans, ammonia, short-chain carboxylic acids, or other odorous or toxic products and are a common cause of biofilter medium replacement.

While the slow diffusion of oxygen in the water or biofilm phase can limit rates of aerobic degradation, it may be exploited in specific cases. Some compounds, especially chlorinated hydrocarbons, can only be degraded under

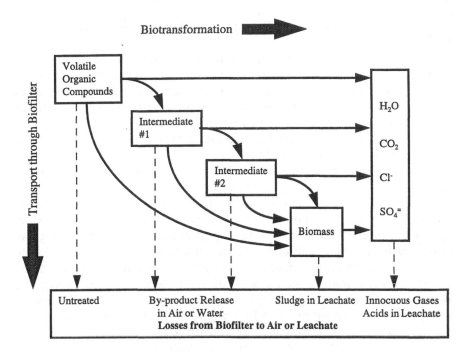

Figure 2.4 Biotransformation and transport processes in biofilters.

anaerobic conditions. Investigators have demonstrated denitrification and reductive dehalogenation in aerobic biofilters (see Chapter 5). Because these processes only occur in the absence of oxygen, it is presumed that they were done by microorganisms in anaerobic regions beneath the aerobic biofilm. Future work may expand the range of compounds susceptible to anaerobic biofiltration.

2.6   *Product generation*

The contaminants entering a biofilter may be energy-rich volatile organics or simpler inorganic compounds such as hydrogen sulfide or ammonia. They might be converted to carbon dioxide, water, or sulfate and nitrate by a single organism (Figure 2.4). Alternatively, the compound could be converted to a secondary product and passed to another organism. A complex or difficult-to-degrade compound might undergo several different transformations in several microbial species before mineralization. An intermediate compound with a high vapor pressure may escape to the effluent air, while one resistant to degradation can accumulate in the biofilter or wash away in leachate. Fucich et al. (1997) studied biofiltration of carbon disulfide. They noted that an intermediate product, carbonyl sulfide (COS) can escape from the biofilter if it is not operated properly. Devinny and Hodge (1995) saw production of ethyl acetate and acetic acid in biofilters overloaded with ethanol.

Some of the carbon in the contaminant will be incorporated into the biomass. The biomass itself will be degraded as organisms die and are consumed by their fellows. Biomass may escape in the air or leachate (Ottengraf and Konings, 1991), but in amounts which are small in comparison to the amounts present in the biofilter. The reservoirs of transformation products (intermediates, biomass, and inorganics) will be adsorbed to the medium or dissolved in the water.

At steady state, the reservoirs will be neither growing nor declining. In a successful low-load biofilter, the biological transformations will be rapid in comparison to physical transport, so that the elements present in the incoming contaminant will be dominantly converted to mineral products. Maintaining the biomass at steady state is particularly important: if it grows continuously, it will eventually clog the biofilter. In some cases, biofilter operators have limited biomass growth by keeping nutrient concentrations low. However, in many systems, particularly those treating high concentrations of readily degradable compounds, significant growth of biomass is anticipated, and a means of controlling pressure drop across the bed, such as medium mixing or repacking, must be employed.

While compounds are transformed, chemical elements are neither created nor destroyed in biofilters, so a mass balance can be calculated. The amount of carbon entering the biofilter, for example, must equal the amount accumulating plus the amount leaving. This provides the means for a relatively rapid and accurate measurement of biomass accumulation: if some of the carbon entering the biofilter is not coming out, it must be accumulating as biomass or by-products. Bench-scale testing of a biofilter to determine whether it will be effective for a given effluent must be done within a few months to be economically feasible. However, biomass accumulation which clogs the biofilter in six months or a year can be a serious operational problem, and it is desirable that the shorter test predict such problems. A carbon balance measurement can provide the needed prediction. The difference between the amount of carbon entering the biofilter and the amount leaving can show the rate of biomass accumulation, and indicate whether this will be a problem in full-scale operation (Medina et al., 1995b; Auria et al., 1996). However, the multiple fates of the carbon must be considered. In experiments with methyl ethyl ketone, Deshusses (1997) found 82% of the carbon which disappeared as treated contaminant appeared as carbon dioxide. Growth of biomass was not sufficient to account for the remainder, and he postulated that carbonate might be lost with leachate.

In a multi-step transformation, establishment of a steady state requires that each of the transformations is occurring at the same rate. If the input exceeds the output for any reservoir, the concentration of that compound will increase. In a study of ethanol biofiltration, for example, a very high feed rate caused accumulation of intermediates and biomass (Devinny and Hodge,

1995). Because one of the intermediates was acetic acid, its accumulation caused a drop in the pH of the medium. This further slowed metabolism, which allowed accumulation of more acid, creating a worsening cycle, and eventually the biofilter failed. At more moderate loads, acidification did not occur, but biomass rapidly accumulated and clogged the biofilter.

2.7 Heat generation

The oxidation of organic compounds also generates heat. While it is commonly said that the microorganisms "use" energy, it is of course conserved, as they convert chemical energy to heat. For some contaminants, the amount of heat is significant. For example, the complete oxidation of 1 g of ethanol in a cubic meter of dry air is sufficient to raise the temperature from 20°C to 38°C. Because temperature is easily measured with great accuracy, it is possible to use heat generation as a measure of degradative activity. However, several factors must be taken into account in order to calculate a heat balance.

In a system which is not at steady state, some of the energy released by biodegradation will contribute to increasing the temperature of the biofilter. The medium and the biofilter vessel absorb energy in amounts which can be estimated from their mass and specific heats and the temperature change. If the biofilter is at steady state so that the temperatures of the medium and the vessel do not change, this factor can be ignored. However, if the biofilter is at a different temperature than its surroundings, heat will be lost or gained through the walls. In experiments, this can be minimized by careful insulation of the biofilter. In large biofilters where the exterior surface-to-volume ratio is small, heat transfer through the walls may also be negligible. With these factors controlled, biologically generated heat will appear in two forms. Water will be evaporated from the medium, and the outlet air will be at a higher temperature than the inlet air.

If the temperature of the air rises as it passes through the biofilter, and the inlet and outlet air are both at 100% humidity, evaporation will be occurring because the water content of warmer air is higher. If the inlet air has a relative humidity less than 100%, the amount of water required to bring it to 100% must also be included. Water contents for air at various temperatures can be found in standard tables (Appendix D), so the amount of heat consumed by evaporation can be calculated from the heat of evaporation of water.

The heat capacity of air can be used to calculate the amount of energy consumed in warming the flow. The sum of the heat of evaporation and heat required for air warming is the amount of biodegradation heat generated. Finally, it can be compared with the heat of combustion of the compound being treated to determine how much of the contaminant is being degraded. The difference between the amount being degraded and the amount represented by CO_2 release from the biofilter gives an indication of how much carbon is being incorporated in biomass (Medina et al., 1995a). Thus, a heat balance calculation can give an independent estimate for the rate of biomass accumulation in the biofilter. This may be far more convenient and accurate than direct measurement of biomass accumulation.

2.8   Conclusions

Biofilters, like all systems, follow the laws of energy conservation and mass balance. If researchers measure the amount of energy entering a biofilter in the form of chemical potential and heat, it can be matched with the energy leaving the biofilter in the form of more lower potential products and more heat. The carbon, sulfur, chlorine, and other elements which enter the biofilter must either accumulate in reservoirs in the biofilter or escape as gases or solutes in the leachate. Following the energy and mass balances can provide better insight into the complex phenomena within the biofilter, allowing operators and designers to maximize the transformations of undesirable pollutants to safe products (and generally maximizing conversion of chemical energy to heat).

The principles which govern phenomena within the biofilter are also superficially quite simple. Compounds partition from the air to the water to the extent given by Henry's Law and diffuse through the water at rates described by Fick's Law, and their adsorption on the medium might be described by Langmuir or Freundlich equations. Monod has given us a relation which describes much about how microorganisms consume the contaminants and grow. However, the irregular shapes of the media, and particularly the overwhelming numbers and diversity of organisms in the biofilter, combine to produce systems which are extremely complex. Even as we recognize general principles, the details force us to admit that we cannot reliably predict the performance of a particular biofilter on a particular waste until we have tested it.

chapter three

Biofilter media

3.1 Introduction

Biofilters work by using a porous solid medium to support microorganisms and give them access to the contaminants in the air flow. The media used vary tremendously. The particles may be organic materials, natural inorganic solids, or entirely synthetic. They vary in size, which affects crucial medium characteristics such as resistance to air flow and total biofilm surface area. The particles may be smooth, with a low surface-to-volume ratio, or they may have complex shapes and internal micropores which create a huge surface area for adsorption. Some media come with a culture of microorganisms, while others must be inoculated. Costs vary tremendously, from a few dollars per m³ for composted yard waste to many hundreds of dollars per m³ for manufactured media or coated polyurethane foam. The nature of the biofilter medium, or packing material, is a fundamental factor for successful application of biofilters. It will affect the frequency at which the medium is replaced and will have a major impact on key factors such as bacterial activity and pressure drop across the reactor. Consequently, the nature of the medium will influence both the removal performance and the operational costs.

From the time biofilters entered the market in the 1970s, researchers have tried to develop better packing mixtures. The result is that there are probably as many different biofilter media as there are biofilters. Advanced packing materials involve complex blending, sometimes with proprietary agents, whereas other low-tech biofilters are simply packed with soils. Most biofilter media include various proportions of biological residues (compost, peat, soil) and inert bulking agents (wood chips, activated carbon, polystyrene beads), supplied with an appropriate bacterial inoculum, pH buffer, and mineral salts. Other materials have also been used in laboratory studies but have not yet been applied in the field (see some of the examples of packing in Section 3.4). Criteria for a good packing are reviewed below.

3.2   Criteria for the choice of an optimal biofilter medium

The choice of materials for biofiltration should be based on the following criteria.

3.2.1   Inorganic nutrient content

High nitrogen, phosphorous, potassium, and sulfate contents, as well as trace elements, are required for the establishment of a dense process culture. At this time, little information exists on nutrient cycles and nutrient requirements in biofilters. In general, nutrients are supplied as slow-release nutrient granules or sprayed as solution onto the medium during initial medium preparation only. However, in some cases, nutrients are added subsequently on a regular basis during operation (see Chapter 4 for a more detailed discussion). In general, for a compost-based medium, an initial addition of N, P, and K in the range of 0.4, 0.15, and 0.15% by weight based on dry packing is considered sufficient.

3.2.2   Organic content

In many cases, biofilters will be exposed to discontinuous emissions of pollutants because of either process rotation or weekend shut-downs. Inert biological residues and other assimilable organics present (for example, in compost) seem to constitute an alternative food source for biofilter microorganisms that can be used during shut-downs.

3.2.3   Chemical and inert additives

Inert additives generally serve multiple purposes. One of their main functions is to prevent compaction in biofilter beds and minimize the pressure drop. Large-size inert additives such as expanded polystyrene beads (e.g., BIOTON® packing; ClairTech BV, Woudenberg, The Netherlands), wood chips (2 to 5 cm in length), bark, expanded clay, glass beads, perlite, vermiculite, tire scraps, etc. have been utilized for this purpose. They are often called bulking agents. Other inert additives serve as a buffer to attenuate fluctuating inlet concentrations. Granular activated carbon (GAC) is probably the most widely used additive for this purpose (Weber and Hartmans, 1995). Another function of additives is to maintain optimum pH and nutrient conditions throughout the reactor. Limestone or crushed shells have been used to buffer produced acidity (Yang and Allen, 1994), and slow-release nutrient granules have been used to maintain appropriate levels of nutrients.

3.2.4 Water content

The medium should have characteristics which aid the operators in keeping the water content at levels which nurture the microorganisms. It should store large amounts of water and make it readily available during periods of drying. This can be described precisely in terms of the "characteristic curve", which relates the water availability on the surface of the medium to the water content over the range from complete dryness to saturation (Chapter 4). In general, however, it is desirable to have media with a high water-holding capacity, and typical organic media may be 40 to 80% water (by weight) when they are saturated.

Some composts which are initially hydrophilic can become hydrophobic when dried. This makes rewetting difficult, so recovery from an inadvertent drying episode is slow or even impossible. The phenomenon has not been studied carefully, and it is not yet known how to choose materials which do not have this characteristic.

3.2.5 pH

For the greatest spectrum of bacterial activity, a near-neutral pH is required. The usual pH value for packing materials is 6 to 8, although in some cases, as when treating reduced sulfur compounds, a pH as low as 2 to 4 has been observed (Furusawa et al., 1984; Webster et al., 1996) without important loss of pollutant removal performance.

3.2.6 Sorption characteristics, porosity

Sorption of pollutant onto the packing includes absorption in the pore water and the biofilm and physical adsorption on the medium matrix. It is a function of the moisture content, the pollutant undergoing treatment, and the nature of the medium. Because the availability of the pollutant to the process culture and sorption are closely related, an optimum biofilter medium should provide sufficient sorption capacity for the pollutant undergoing treatment. In the event of fluctuating inlet conditions, sorption of the pollutant onto the medium will play an important role in attenuating concentration peaks.

In most cases, a homogeneous filter bed with a porosity or void volume of 40 to 80% will ensure both gas plug flow and a low pressure drop. Packing should guarantee a large surface area for both microbial immobilization and pollutant mass transfer. Hence, bed particles should be relatively small (diameter of 1 to 5 cm).

3.2.7   Bacterial attachment

The support material should have suitable properties for bacterial attachment. Research has indicated that media which are rough, porous, and hydrophilic are more readily colonized by microorganisms (Durham et al., 1994). The rough surface provides niches that protect the organisms from any hydraulic shear. Pores that are large enough to hold microorganisms may become refuges in which the microorganisms survive adverse conditions, ready to recolonize the surface when conditions improve. Presumably, the substances which microorganisms exude in order to form attachments to the surface stick more readily to hydrophilic surfaces. With synthetic media intended for use chemical scrubbers, care should be taken because antimicrobial agents or coating may have been added to the media to prevent slime growth.

3.2.8   Mechanical properties

The filter bed structure should remain stable with time. No clogging or shrinking of the bed due to material decomposition, bed compaction, or water condensation should occur. This puts constraints on the density of the medium in relationship to its structural strength. The material at the bottom of the bed must bear the weight of the material above it. A medium which is heavy and soft will compact at the bottom if the layer is too deep. Compost beds have a lower density of 300 to 500 kg m^{-3} (wet) but are easily compacted and so are limited to layers about 1 to 1.5 m deep. Soil beds are generally much more dense, at 1000 to 5000 kg m^{-3}, but they are also quite resistant to crushing and so can be made deeper. Inorganic materials such as activated carbon can be piled deeper, and plastic shapes may be stacked as deep as 5 m.

Greater depth is desirable because it means a larger volume reactor can be placed in a smaller space without having to build expensive multi-layer biofilters. Lower density is desirable because it means the reactor vessel can be lighter and less expensive. This is especially important if multiple trays or stacked biofilters are used; light weight is valuable when a more complex structural support is required.

3.2.9   Odor of the packing

Because the packing is made of organic material and supports biomass, it has an intrinsic smell: the usual values for effluent air are 20 to 100 odor units per cubic meter (OU m^{-3}). This odor is generally perceived as the pleasant odor of a forest after a rainy day. Even so, biofilters are not recommended for the treatment of indoor air in closed loop.

3.2.10 Packing cost and lifetime

The packing material should provide good removal characteristics over a period of 2 to 4 years. Its price should be small compared to the other investment costs, and its contribution to the overall operating cost should remain minimal. Medium replacement is usually necessary when either removal performance drops below the acceptable limit or when pressure drop becomes too high and no remedial action proves useful in restoring optimum conditions. In some cases, medium lifetime can be extended by breaking the surface crusts or removing the medium from the reactor and mixing it in order to restore its original structure. Typically, soil media are the least expensive and may be available for the trucking cost. Compost-based media are more expensive, anywhere from $50 to $500 per cubic meter of medium depending on the vendor, the delivery location, and the quantity needed. Synthetic media are even more expensive.

3.2.11 Packing disposal

The used packing material should not be an hazardous waste and should be easily and cheaply disposable, e.g., by land farming. If required, the separation of the bulking agent from the peat or compost should be easily achieved.

3.3 Materials used for biofilter media

A large number of biofilter media have been used either in laboratory-scale experiments or in the field. The individual components are listed below and their properties discussed. A summary of important properties for selected materials is given in Table 3.1, and typical biofilter media are described in Section 3.4.

3.3.1 Compost

Finished compost possesses a large diversity and density of microorganisms. It has good water retention properties, neutral pH, and a suitable organic content. Its pressure drop is generally higher than peat, and it is subject to bed compaction; therefore, it is usually mixed with various proportions (20 to 80%) of bulking agents (wood chips, perlite, etc.). Several composts have been used, such as sewage sludge, yard waste, or manure composts. Every manufacturer chooses a different compost, but studies do not indicate that the biofilter performance is greatly affected by the type of compost. Since the mid-1990s, the use of pelletized compost media (Oude Luttighuis, 1997; Proell et al., 1997) has been attempted. Such media showed low pressure drop compared to traditional compost beds. This development still remains experimental.

3.3.2 Peat

Peat is naturally acidic and hydrophobic. Because of its hydrophobicity, moisture control for peat beds can be difficult. Peat does not naturally

Table 3.1   Summary of Important Properties of Common Biofilter Materials

	Compost	Peat	Soil	Activated carbon, perlite, and other inert materials	Synthetic material
Indigenous microorganisms population density	High	Medium–low	High	None	None
Surface area	Medium	High	Low–medium	High	High
Air permeability	Medium	High	Low	Medium–high	Very high
Assimilable nutrient content	High	Medium–high	High	None	None
Pollutant sorption capacity	Medium	Medium	Medium	Low–high[a]	None to high[c], very high[a]
Lifetime	2–4 years	2–4 years	>30 years[b]	>5 years	>15 years
Cost	Low	Low	Very low	Medium–high[a]	Very high
General applicability	Easy, cost effective	Medium, water control problems	Easy, low-activity biofilters	Needs nutrient, may be expensive[a]	Prototype only or biotrickling filters

[a] Activated carbon.

[b] Bohn (1988, 1996).

[c] Synthetics coated with activated carbon.

contain a large population of microorganisms and will require inoculation, e.g., with activated sludge. It has much less nutrient than compost and thus may require nutrient supply, e.g., either through initial addition of slow release nutrients or trickling of a nutrient solution during operation. Peat was widely used as a medium in the 1980s because it offered a very low pressure drop. It has been slowly replaced by compost/bulking agent mixtures, mostly because of the better performance of the latter medium in the long run and because of the relative difficulty of controlling moisture in peat beds.

3.3.3 Soil

Soil has been utilized as a biofilter medium because it is inexpensive and plentiful and has a large indigenous microbial population. The first biofilters were constructed by digging a large pit in the ground, inserting air distribution pipes at the base, and filling the pit with soil (Pomeroy, 1957). Soil as a filter medium has been extensively studied by Bohn (1975, 1988, 1996). Soils are naturally hydrophilic and are less difficult to rehydrate than compost or peat in the event of inadvertent drying. Soils do not have the tendency to aggregate, but their permeability remains low; therefore, soil beds have large pressure drops and often develop preferential paths for air flow, isolating portions of the biofilter. Another drawback is their low specific activities (Bohn, 1996). This means that larger reactors are required, and soil beds are generally considered where space is not limiting. However, soil has a high bearing strength, so soil can be layered with much less compaction than compost or peat, usually without structural support. Soil is best used for low-tech, open-bed biofilters. Because of their low permeability, soil biofilters are usually operated with large gas residence times (minutes).

3.3.4 Activated carbon

Granular activated carbon (GAC) has been used as the biofilter medium in several studies (Weber and Hartmans, 1995; Graham, 1996; Webster et al., 1996). For example, Wheelabrator Water Technologies, Inc., has used granular carbon biofilters successfully for full-scale treatment of petroleum hydrocarbons from a refinery and from a soil vapor extraction project (Graham, 1996). GAC can be used either alone (Graham, 1996) or as bulking agent mainly to attenuate pollutant fluctuations (Weber and Hartmans, 1995). It can be provided in any desired particle size, but coconut shell carbon in cylinders 0.3 cm in diameter by 0.5 cm long has commonly been used. Granular activated carbon has excellent structural properties, with uniform particle size and good resistance to crushing. It has substantial water-holding capacity and provides a good surface for microbial attachment. While its adsorptive capacity in a biofilter is reduced by the water and biomass on its surface, it remains much higher than other media. Activated carbon must be prepared for use in a biofilter. Nutrient amendment and microbial inoculation are required, and some carbons are strongly basic and must be neutralized to provide a suitable pH. Activated carbon has worked quite well in biofiltration of hydrocarbon fuel and ethanol vapors. Its primary disadvantage is its high cost (approximately $900 to $1000 m^{-3}), but it is expected to last many years.

3.3.5 Wood chips or bark

Wood chips or bark are commonly used in various proportions mostly as bulking agents, but there are several reports on the use of wood chips alone as medium (Paul, 1994; Finn and Spencer, 1997). In the latter case, regular nutrient supply is needed. Common particle sizes are 1 to 5 cm. In addition to preventing bed compaction and allowing for homogeneous air flow, wood chips or bark constitute a reservoir of water that may in some case attenuate fluctuations in packing moisture content due to poor reactor control or excessive heat generation. So far, there is no report indicating that a particular species of tree would be more suited for biofiltration purposes; however, it is reasonable to assume that some may be better than others. Many barks and woods contain antibiotic substances that the tree has synthesized to protect itself from disease and rot. While no research has been done on the issue, it may be that some of these could interfere with biofilter operation.

3.3.6 Perlite

Perlite is a very light porous material available in different sizes (3 to 15 mm in diameter). It is inexpensive and has a large surface area, but it contains no nutrients and no microorganisms. It has been used as a bulking agent (Shareefdeen et al., 1993) or as sole medium in biofilter experiments (Cox et al., 1997).

3.3.7 Synthetic media

Various synthetic packing materials have been tested for biofiltration, mainly at the bench scale. Synthetic materials do not contain nutrients or microorganisms, so these must be added. As in any biofilter, nutrients may be lost to the leachate or sequestered in the biomass during operation. Because there is no release of nutrients from the media, as occurs with slowly decomposing compost, they must be resupplied by continuous or occasional additions in the irrigation water. The control of water and nutrient supply depends on the specific nutrient requirements, on the water sorption properties of the support, and on the need to control and wash out produced metabolic acidic byproducts. Activated carbon or extruded diatomaceous earth are known to be porous enough to hold significant amounts of water and nutrient; hence, the water content in reactors packed with these media generally can be easily controlled. Synthetic media such as polypropylene rings have no water-holding capacity and will require intensive trickling to keep surfaces wet. In the long run, uncontrolled growth of biomass due to continuous nutrient supply can become a major problem in synthetic media biofilters. Innovative research with inert support materials is generally not reported in order to protect potential patents. Materials such as vermiculite (Ortíz et al., 1998), silica, cordierite, or ceramic monoliths coated or not with activated carbon (Bishop and Govind 1995; Sorial et al., 1995), polyurethane foam (Cox et al., 1993; Loy et al., 1997; Moe and Irvine, 1997), extruded diatomaceous earth (Kinney et al., 1996a), ceramics (van Groenestijn et al., 1995), polystyrene coated with powder activated carbon (De Filippi et al., 1993), pelletized synthetics (Sorial et al., 1995), glass wool, or other fiber materials have been reported. So far, in most cases, the moderate performance improvement over traditional compost beds does not justify the additional costs of synthetic media.

3.4   Description of selected biofilter media

The following selection of media is not an endorsement of any specific medium but is intended to aid the reader in understanding the many formulations of biofilter media.

3.4.1   Compost-based and organic media

Various mixtures of peat, perlite, vermiculite, and shredded polyurethane foam were tested in New Jersey Institute of Technology laboratories (Shareefdeen et al., 1993). The best results with respect to performance and long-term stability were obtained with a mixture of peat moss (Hyponex; Marysville, OH) and perlite (Grace & Co.; Cambridge, MA) at a 2:3 volume ratio (before mixing). The addition of perlite was an important factor for reducing the pressure drop in peat moss beds (Shareefdeen et al., 1993).

A mixed medium was evaluated for biofiltration purposes at the University of California, Riverside. The medium included 80% by volume of wood chips (1 to 5 cm in length) and 20% compost (Deshusses et al., 1997). It was amended with 0.4, 0.15, and 0.15% (dry weight basis) N, P, and K, respectively, and buffered with about 25 kg of finely crushed oyster shells per cubic meter of medium. The finished medium weighed about 430 kg per cubic meter of bed.

Various supports were tested at the Universidad Autónoma Metro-politana-Iztapalapa, (UAM) in Mexico City (Ortíz et al., 1998). These included mixtures of peat, vermiculite, porous ceramics, and pine bark. Among those tested, a mixture of two thirds (by volume) vermiculite and one third activated carbon presented the best results for BTX vapor removal. It was found that vermiculite presented favorable properties for biofiltration, such as a large interfacial area (about 650 $m^2 \, m^{-3}$), low specific gravity, relatively homogenous particle size distribution, and good water drainage. This resulted in biofilters with a relatively low pressure drop (Ortiz et al., 1998).

The BIOTON® medium is a patented mixture composed of approximately equal volumes of polystyrene beads (4 to 8 mm in diameter) and proprietary compost, mixed in a proprietary process with nutrients and pH buffer. The BIOTON® medium weighs 300 to 400 kg m^{-3}, and its optimal moisture content is approximately 60%.

The medium used by The Reynolds Group is composed of about 20% redwood chips (1 to 5 cm length), 50% finished horse manure compost, and 30% coarse yard waste compost. The compost is specially prepared for biofiltration. A first sieving to remove particles smaller than 6 mm is performed, the material is allowed to compost for 2 to 4 months to remove any easily biodegradation organics, and the finished compost is sieved again to remove fine particles smaller than 6 mm. The medium is also amended with nutrient and lightly buffered.

3.4.2   Soil media

Bohn Biofilter Corporation uses soil. In general, to save on medium transportation costs, appropriate soil(s) are sought within reasonable trucking distance of the biofilter. Bohn's methods to select the most appropriate soils for biofiltration remain proprietary information.

3.4.3   Synthetic media

Cox at TNO developed a packing made of perlite (Cox et al., 1997). Prior to use in the biofilter, the perlite (mean diameter 4.5 mm) was saturated with a mineral medium and seeded with an enrichment culture. During biofilter operation, mineral medium was added to the top of the biofilters at regular intervals (once every 2 to 3 weeks) to maintain optimum moisture and nutrient content.

Devinny and coworkers completed several experimental studies using activated carbon as a biofilter medium (Medina et al., 1995a,b; Webster et al., 1996). They chose coconut carbon in 3-mm diameter cylinders. This was submerged in water and neutralized to pH 7. Plant fertilizer was added with an inoculum of microorganisms from several sources, including sewage sludge, soil, compost, and previously used biofilter medium. The suspension was mixed, then drained, and the wet carbon was placed in the biofilter vessel.

chapter four

Controlling factors and operation of biofilters

4.1 Introduction

Many physico-chemical and operational factors influence performance, treatment costs, and long-term stability of biofilters for air pollution control. In this chapter, emphasis is placed on identifying factors that affect biofiltration, explaining and discussing their influence, and providing guidelines on how to control these factors to optimize operation. In general, the three most important parameters for an efficient biofilter are medium moisture content, pH, and bed temperature. Other factors are also important, but they influence medium lifetime or removal performance to a lesser extent than do these three factors. The fundamental means of treatment in biofilters is the action of pollutant-degrading microorganisms. This means that controlling operating parameters in a biofilter is an attempt to control the activity of the process culture. This is complex, and the subtleties of understanding and difficulties of control should not be underestimated.

4.2 Water content

The amount of water in a biofilter is perhaps the most important parameter under the control of the operator. Neglect of the water content or difficulties in controlling it are the most common cause of poor biofilter operation. Biofilters for air pollution control differ from those used for water treatment because most of the pore space is filled with air, but water remains a necessity: microorganisms cannot be active without it. Its presence affects the transfer of contaminant from the air and the physical properties of the medium.

The principles which govern the behavior of water in porous media are not generally taught to engineers. Soil scientists, in contrast, have studied how water and soil particles interact and how they influence the soil organisms for many years. Their knowledge is immediately relevant to biofiltration, and should be studied by biofiltration engineers.

4.2.1   Water in porous media

Many solids have a much stronger affinity for water than air does. An individual molecule of water will stick to the surface of inorganic minerals with great force. Air is a non-polar, hydrophobic medium, while most solids have at least some dipole moment to attract the polar water molecules. In the case of clay minerals, for example, the surfaces are charged and will attract water molecules strongly. Some materials, such as peat, are hydrophobic and are not easily wetted. Others yet may change with conditions. The common observation of difficulties in rewetting dried biofilter medium suggests that the initially hydrophilic compost becomes hydrophobic when it has been dried.

The interactions of water with porous media can best be summarized by describing what happens as water is added to a dry medium. As a dry porous medium encounters water vapor, it will have a strong tendency to draw water molecules out of the air. In this regime, water exists as individual molecules adsorbed to the soil surface. Within such dry material, the water activity (or water chemical potential) will be quite low (Figure 4.1).

Before the surface is covered, the water activity is primarily controlled by the strength of attraction of the water molecules to adsorption sites on the bare surface. As adsorption sites become less abundant, it becomes less likely that a water molecule will be adsorbed and more likely that one will be released, so the equilibrium relative humidity rises. But, in this regime, humidities are well below 100% and the environment is very unfriendly to proliferation of most microorganisms. Completion of the monolayer occurs at water levels below 1% of field capacity in most soils (Unger et al., 1996). Water contents that are not sufficient to complete a monolayer will not be of concern in understanding biofilters. Bacteria cannot be active in an environment where water exists only as individual molecules adsorbed to a solid surface.

The amount of water required to complete a monolayer varies with the surface area of the medium. In fine clays, which have very high surface areas per unit volume, a monolayer of water may represent a water content of 10% by weight (Devinny, 1989). In large-particle sands with low specific surface areas, a monolayer of water will constitute much less than 1% by weight. This difference in surface area is an important contributor to the fact that the relations among water activity, soil moisture content, relative humidity, and biological availability of water will be very different in soils with different physical characteristics.

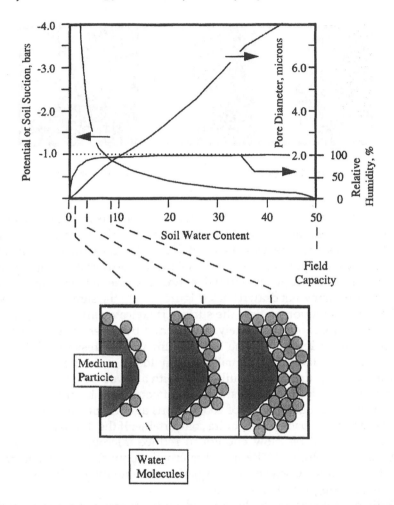

Figure 4.1 Water activity, soil suction, and threshold pore diameter in porous media.

After the monolayer is complete, further addition of water will place water molecules on other water molecules rather than on the medium surface. This is a substantial change. The humidity over a pool of pure water, which is at equilibrium with water molecules sticking to other water molecules, is 100%. Within soils and other porous media, there are two reasons why humidity does not rise to exactly 100% when the second layer of water molecules is begun. First, the initial layer of water molecules does not entirely shield the second layer from the effects of the support material. A polar mineral will tend to polarize the water molecules adsorbed to it. A molecule in the second layer thus "sees" a water layer which is more attractive than bulk water. The surface attraction can propagate through several water layers if it is strong.

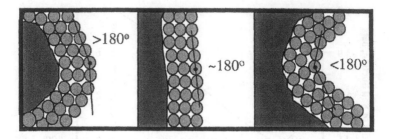

Figure 4.2   Effects of surface curvature on water films.

Second, the water layers on the surface of a porous medium are not flat, like the surface of a pool of water. Each molecule at the surface is exposed to attractive forces from the water below it and to the sides, but not from above. But, if the water layer is convex, as it is on the surface of a round particle, molecules are exposed to less attractive force because the water molecules at their side are somewhat drawn back (Figure 4.2). In the same way, molecules in a layer within a pore, where the surface is concave, are more tightly held because they are more completely surrounded by other water molecules.

This in turn determines the equilibrium vapor pressure of the water. More concave surfaces with a small radius hold the molecules firmly, reduce the water activity, and are at equilibrium with a lower water vapor pressure. While this effect seems rather subtle, its effects are substantial. A concave water surface with a diameter of 100 microns at equilibrium has a potential of –0.03 bar, where bacteria are quite comfortable. If the diameter is 1 micron, the potential is –3 bar, low enough to reduce microbial respiration rates substantially (Stolp, 1988). Because 1 micron is about the size of a bacterium, this is the potential at which microorganisms can no longer find water-filled pores large enough for them to enter.

Individual water molecules are constantly evaporating and recondensing. In time, all of the water in a small volume of porous medium will be at equilibrium with the same vapor pressure. Correspondingly, all of the concave water surfaces will have the same radius of curvature. If the radii of the water surfaces are different, water will tend to evaporate from the pore with the larger radius and condense where the radius is smaller, until large pores are empty and small ones are filled (Figure 4.3).

If the pore itself has a radius of curvature smaller than the thermodynamic equilibrium value for the water, it will be filled with water. If the pore radius is larger than the equilibrium water value, it will be empty except for the thin layers held on the surface by adsorption. Thus the equilibrium water radius, r_w, is also the "threshold" pore radius. Pores with radii smaller than the threshold radius will be filled with water, and larger pores will be empty. This threshold is an important operating parameter for a biofilter. If too many large pores are filled, anaerobic conditions may develop, and air flow

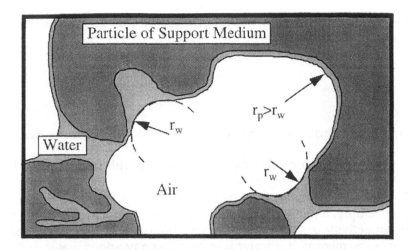

Figure 4.3 Distribution of water in porous media.

will be hindered. If only the tiniest pores hold water, microbial activity will be inhibited. In between, there is a range of relative humidities that will provide appropriate conditions.

4.2.2 Thermodynamic relationships

The strength with which the molecules are held in the soil water by the mineral surface or by other water molecules is measured either as the chemical potential or as the water activity. These two quantities are directly related, (Stolp, 1988):

$$\psi = \frac{RT \ln a_w}{10V_w} \quad or \quad \ln a_w = \frac{10V_w \psi}{RT} \tag{4.1}$$

where ψ = water potential or soil suction, bars; R = gas constant (8.314 J mol^{-1} K^{-1}); a_w = water activity = relative humidity × 10^{-2}; V_w = partial molal volume of pure water, 0.018 L mol^{-1}.

The water activity is the idealized thermodynamic concentration and is directly related to relative humidity. Pure water is assigned activity 1.0 (potential = 0), and the air over it is at a relative humidity of 100%. Water tightly held by attraction to the porous medium surface has a potential below zero, because energy is required to move it to a body of pure water. Accordingly, its activity is below one, and the relative humidity is below 100%. In contact with water, the soil will "soak it up" like a dry sponge. Thus, the potential is often call "soil suction" or "soil water tension" and can be measured in bars or other units of pressure.

The relationship between water potential and the curvature of the water surface in pores is also known:

$$r_w = \frac{20\gamma}{\psi} \tag{4.2}$$

where r_w = equilibrium water radius, in cm; γ = the water surface tension, in Newtons per cm.

This value is also the threshold pore radius, with smaller pores being filled and large pores being empty.

For perfectly dry materials, the water activity is very low, the smallest pores are empty, and the relative humidity is zero. At a high water content, the water activity is near 1.0, the relative humidity is near 100%, and the affinity of the medium for water is so low that any added water drains off under the force of gravity. At this second point, the medium is said to be at "field capacity", and biofilters are typically operated with water contents at or near this value. Biotrickling filters are operated at water contents well above field capacity, so drainage is continuous.

Between dryness and field capacity, the combination of surface composition, surface area, and pore size distribution gives each porous material its own "characteristic curve". This curve relates water content to water potential. A hypothetical characteristic curve is shown in Figure 4.1, with relative humidity and the threshold pore diameter displayed as a function of water content. The water potential rises to high (negative) values when the soil is very dry. It is zero when the soil reaches field capacity, the point where soil suction becomes weaker than the force of gravity. The moisture content at field capacity varies greatly for different soils or biofilter media, because of their different surface areas, affinities for water, and pore-size distributions. The shape of the characteristic curve is equally variable. Both of these must be measured individually for each medium. The threshold pore diameter increases as the water content rises, so that the pores in the support medium are filled sequentially from the smallest to the largest.

4.2.3   Biological effects

The range of water activities found in soils and biofilters defines a wide range of ecological regimes (Stolp, 1988). Bacteria of the genus *Spirillum* require water activities very near 1.0. *Pseudomonas fluorescens* does well at 0.97, and the common soil bacterium *Aerobacter aerogenes* can survive at 0.94. Activities between 0.8 and 0.9, however, support only bacteria specialized for survival in dry conditions, and below 0.8 systems are dominated by xerophytic fungi.

Microorganisms cannot survive where the soil suction exceeds their tolerance limits. Even within the range where they can survive, a greater suction means they must expend greater energy to draw water into the cell, and growth will be reduced. The most biologically relevant parameter describing water content is the chemical potential, because it is a direct measure of how much energy the organisms must expend to collect water. A given moisture content will be associated with very different potentials in different media.

Water serves as the medium within microbial cells, dissolving the myriad chemicals which constitute the life of the microorganism. Each cell must control its water content. In pure water, cells face the problem of osmotic pressure: water tends to flow into the cell because the high concentrations of solutes make the water activity there lower. On the surface of particulate media, however, a microorganisms may have difficulty getting water into the cell.

The metabolic activity of the cell tends to deplete substrate and increase the concentration of waste in its immediate neighborhood. As concentration gradients are established, substrate will diffuse toward the cell, and waste materials will diffuse away. Diffusion in water is slower than in air, commonly by a factor of 10,000 (Cussler, 1997). Thus, a cell at the bottom of a deep pore filled with water or at the base of a biofilm may have difficulty obtaining substrate and oxygen from the air and must face higher concentrations of its waste products.

Unlike fixed-film reactors used for water treatment, gas-phase reactors may include a film of standing water outside the biofilm. This could be the case if the water content of the biofilter is increasing, so that the thickness of the water film is growing faster that the thickness of the biofilm. Microorganisms can swim in this water, and biofilters can maintain communities of free-swimming organisms which are not seen in other biological reactors. If condensation or irrigation is heavy, so that some leaching of water is occurring, these microorganisms can be carried out of the biofilter.

4.2.4 Partition effects

The first step in biofiltration is transfer of contaminants from the air to the water. Unger et al. (1996) note that organic compounds in a damp porous medium may be adsorbed in five ways: vapor adsorption onto solid surfaces, vapor condensation in micropores, partitioning into soil organic matter, dissolution into adsorbed water films, and adsorption to the gas-liquid interface. Condensation in micropores only occurs when the soil is very dry and so is unimportant for biofilters. Adsorption on solid surfaces will be reduced by competition from the water, but may be significant for very adsorptive media such as activated carbon. Absorption within organic matter likely occurs for compost media, but compounds in this state may be largely unavailable for biodegradation.

Contaminants can be biodegraded in the water phase, at the water surface, or possibly while adsorbed to the support surface. For highly water-soluble compounds, concentrations in the water phase will be high, and the amount of water present will have an important effect on the amount of contaminant absorbed from the air. More water means more dissolved contaminant, more opportunity for decomposition, and more rapid and effective treatment. This effect has been noted in biofiltration of ethanol (Hodge and Devinny, 1995). Higher water contents in the biofilter increased the retardation factor, retaining more ethanol and improving treatment success.

Unger et al. (1996) have shown in soil experiments that a substantial portion of a volatile organic compound may be adsorbed at the water-air interface. In the case of a medium with low adsorptivity and a contaminant with low solubility, this could have noticeable effects on partitioning of the contaminant from the air to the water phase, and the interfacial area will determine contaminant partition. This may mean that the addition of water actually causes the mass partition coefficient to increase for modest water contents, and the greatest partition will occur at the water content which produces the highest water surface area (not including the monolayer).

Studies of the "leptopel" or water-surface organisms in the ocean show that there are microorganisms which thrive at the surface where hydrophobic compounds are concentrated. The degree of concentration can be surprisingly great. Geyer et al. (1996) reviewed the literature while studying the fate of pesticides in rice paddies and found reports that the concentrations very near the water surface for PCBs, PAHs, and chlorinated pesticides could be enhanced by factors of 2 to 10,000, and that the factor for PCBs with cosolvents can be 10^9. No studies of such phenomena have ever been done in biofilters.

4.2.5   Interference with air flow

Air cannot flow through pores filled with water. In addition, capillary forces cause water to "bridge" across the spaces between medium particles where they are nearly touching (Figure 4.3). Water may soften some media, contributing to compaction. If it flows, it will carry small particles with it, possibly depositing them in masses which can clog sections of the biofilter, the drainage layer, or the piping (Leson, 1993). For all of these reasons, excess water in a biofilter will tend to interfere with the passage of air. For media with large, stable particles, such as activated carbon, the effects are negligible. However, if the medium includes small particles, such as soil and compost, too much water can block air transport to volumes within the biofilter. The resulting anaerobic zones are likely to be ineffective in treating the contaminant. They will also produce odorous products such as hydrogen sulfide and mercaptans. As the water content of the biofilter rises, the water will fill more of the smaller pores, reducing the amount of surface exposed to the atmosphere. This will reduce transfer rates for contaminant and oxygen (Alonso et al., 1997; Swanson and Loehr, 1997).

4.2.6 Drainage

When the water content of a biofilter exceeds the field capacity of the medium, water will drain under the force of gravity. Biotrickling filters are
operated so that this occurs continuously in large amounts, and the drainage
is recycled. But, biofilters may also produce some drainage if they are
overwatered or if a cooler biofilter causes condensation of water from the
incoming air stream. The drainage will contain cells, unconsumed contaminant, contaminant by-products, humic materials, nutrients, acids, and salts.

The contaminants and contaminant by-products in the leachate from
biofilters may require treatment before disposal. Loss of nutrients to the
drainage may mean that nutrient supplements for the biofilter will be necessary (Smet et al., 1996). Where these are the important effects, it is best to
minimize drainage by careful control of irrigation (Chapter 7).

A small amount of drainage may be helpful, however, where it carries off
species that can be harmful to the operation of the biofilter. If a small amount
of hydrochloric acid is produced from the degradation of chlorinated hydrocarbons, or sulfuric acid results from the oxidation of organic sulfides, slow
drainage may reduce acidification of the biofilter. Accumulation of salts may
be reduced in the same manner.

Some newer biofilters have soaker hoses installed within the bed. This
allows water to be introduced to the lower level of an up-flow biofilter,
putting the water where it is most rapidly lost to evaporation. It also allows
application of excess water and production of leachate to be confined to the
zone where most acids and salts are produced. The lowest, most impacted
zone can be washed separately without the drainage passing into other
portions of the biofilter.

4.2.7 Control of water content

If a biofilter receives no direct irrigation, its water content will be controlled
by the humidity of the incoming air stream. But, in the range of moisture
content over which biofilters are typically operated (~50% of field capacity),
the equilibrium relative humidity is usually above 99% (Figure 4.1). The
slope of the humidity-water content curve is very small, so that tiny changes
in humidity are associated with large changes in water content. If the relative
humidity of the incoming air drops 1%, because of a change in air temperature or pressure, the soil moisture may drop by 10% or more (van Lith and
Leson, 1996). Control of medium water content solely through control of the
incoming humidity can thus be difficult and may be impossible because of
temperature increases arising from biological oxidation. Very accurate measurements of humidity are necessary if they are to be used as a gauge of
medium water content.

Fortunately, while a drop of 1% will ultimately cause severe bed drying,
it will not do so quickly. If the air enters at 98.5% relative humidity and leaves

at 99.5%, water is removed from the bed, but the rate of removal is low. van Lith and Leson (1997) have developed a systematic method for calculating bed drying rates, and suggested control methods appropriate for various regimes. Biofilters losing less than 50 g m^{-3}h^{-1} can be watered manually, with bed moisture content determined a few times per year. For evaporation rates of 50 to 180 g m^{-3} h^{-1}, a fixed irrigation system should be installed and operated as indicated by a regular schedule of moisture content determinations. If rates are between 180 and 400 g m^{-3}h^{-1}, the irrigation system should be controlled by an automatic timer which provides water at regular intervals, with the length of the intervals and spray time adjusted according to monitoring results. This allows frequent, evenly spaced waterings that will provide large amounts of water without leachate production. If rates are above 400 g m^{-3}h^{-1}, drying problems will be severe. The medium should be carefully chosen for large particle size and water resistance, and the irrigation system should be operated by automatic sensors that detect low moisture contents.

The complex set of factors which influence water content can be a source of difficulties in the field. Drawing on their experience with biofilters treating gasoline vapors from a soil vapor extraction project, Wright et al. (1997) recommended including ports for sampling the medium for water content determinations, incorporating at least two means of adding water, and shading and insulating the units to minimize temperature fluctuations. Pinnette et al. (1995) provided methods of calculation for many of the factors which influence heat and water balance, including humidification systems, metabolic heat, solar radiation, radiative heat transfer, conductive heat transfer, and rainfall.

4.3   Temperature

Microbial activity and biofilter success are strongly influenced by temperature. A microorganism is a tiny bag of chemicals and enzymes, and life is a myriad of chemical reactions that run faster as the temperature rises. Successful microorganisms must coordinate these reactions. If some run ahead of others, excesses and shortages of compounds will develop, threatening the organism. Each species is adapted to control its reactions within a certain temperature range, synthesizing and activating enzymes as needed to maintain control.

Most reaction rates approximately double when the temperature rises 10°C, and microbial metabolic activity will increase proportionately through the range in which the coordination of the reactions can be maintained. In general, a warmer reactor will treat contaminants more rapidly.

There must be limits, however. High temperature may make some reactions occur so rapidly that metabolic coordination is disrupted. Enzymes are made of proteins. Each enzyme has a temperature limit beyond which it is denatured and is no longer effective. Other cell components, such as the

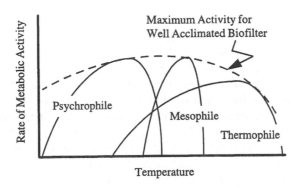

Figure 4.4 Temperature effects on species of microorganisms and biofilter activity.

structural lipids of the membranes, can also be decomposed by high temperatures. As the temperature increases, each microbial species reaches a point where it can no longer be effective, and metabolic activity drops off rapidly. Ultimately, the cell is killed by the heat.

If temperatures decline, the metabolism of the cell will slow, reducing the rate at which treatment proceeds. Eventually microorganisms will become essentially dormant and again may die as necessary functions cease. While some microorganisms are active at temperatures near the freezing point of water, and many can survive freezing, none can be active while frozen.

The combination of these factors produces a temperature-activity curve for a given cell which typically rises with temperature to a limiting value then falls rapidly (Figure 4.4). Different species have different maxima and ranges over which they can survive, reflecting the diversity of natural environments. A species adapted to life in tundra soils might be vigorous at 3°C, damaged by the heat at 15°C, and capable of surviving dormant for long periods while frozen. Species found in hot springs may not grow well below 80°C, and some have been found which can grow near deep sea thermal vents at temperatures over 100°C.

If a biofilter contained only one species, the best strategy would be to operate it at the temperature of the peak of this curve. However, biofilters contain hundreds or thousands of species (Chapter 5), and the long-term response of the microbial ecosystem may be much different than the short-term response of a single species. When an active microbial ecosystem is subjected to a sudden temperature change, many of the species will become inactive, and overall treatment may decline sharply. But, if the temperature change is not too great, much of the treatment effectiveness will be gradually restored. Individual species may become acclimated, those species which can tolerate the change will become active and abundant, and new tolerant species which enter the biofilter will establish themselves (Chapter 5). Water treatment and composting practice has long recognized the utility of psychrophilic, mesophilic, and thermophilic organisms for treating wastes under cold, moderate, and hot conditions.

The effects of temperature are thus strongly dependent on time. A biofilter which operates continuously at high temperatures (say, 60°C) may develop a healthy community of active thermophiles which will rapidly degrade the contaminant. A biofilter treating the same contaminant at 10°C may be slower but still support a healthy community of microorganisms which can make the system successful. Either biofilter, however, could be inhibited by a rapid change in temperature. A sudden change to cold air could reduce treatment rates in the thermophilic biofilter to zero, and a sudden burst of hot air could devastate the psychrophilic microbial ecosystem. Thus, while warm biofiltration is likely to be more efficient than cold, low-temperature treatment may still be practical for many applications. Fluctuating temperatures may cause the greatest difficulties.

While a warmer biofilter generally supports more active organisms, the physico-chemical effects of higher temperatures are usually unfavorable. For most gases, the Henry's Law coefficient rises with temperature, so that less of the compound will be dissolved in the water. Sorption of the compound will probably also be reduced. The net effect is that transfer of the contaminant from the air to the microorganisms will be less effective. It is generally believed that the biological effect is more important than the physical effect, so warm biofilters work better; however, the physical effects should be considered, especially for contaminants with high Henry's Law coefficients.

Unfortunately, control of temperature in biofilters is often limited. Direct heating of the waste air by burning fuel or cooling it by refrigeration would be far too expensive; however, some limited measures are possible. When air is taken from several sources, it can be mixed to provide more moderate and constant temperatures. If the air is too hot, it may be possible to cool it using ambient air in a heat exchanger or by evaporative cooling during humidification. There will be capital costs for the heat exchanger, but no fuel costs. In cold climates, air from a heated facility may be treated in an outdoor biofilter. The air cools as it passes through the ducting to the reactor, and the biofilter cools because heat escapes through its walls. Temperature control in this case may be improved by insulating the ducting and the biofilter and ensuring that the air flow is maintained at all times so that the biofilter is not allowed to freeze. Some flow must be continued at night and through the weekends.

4.4   Medium pH and alkalinity

The effects of pH and temperature on biofilter success are analogous in several ways. Each species of microorganism is most successful over a certain range of pH and will be inhibited or killed if conditions move outside this range. The ranges for different species may be narrow or broad. Some species do well at high pH, and some at low pH, but species tolerant of moderate pH are probably more common. Rapid changes in pH are damaging to most species. Microbial ecosystems, however, adapt to slow changes in pH: species tolerant of the new conditions replace those which are not.

Most biofilters are designed for operation near pH 7. This is generally accepted as a benign condition, and designers are most familiar with it. However, it should be kept in mind that microorganisms are abundant and active in many natural ecosystems where the pH is lower or higher. Alkaline springs are habitats for microorganisms, as are acid bogs. Ideally, treatment of a contaminant should be tested at many values of pH over a wide range in long-term tests in order to determine the global optimum, but this is rarely done because of the cost. The *a priori* assumption that a given compound is best treated at pH 7 is commonly chosen as a conservative guess, but it is often not supported by actual data. Indeed, some systems are successful at other pHs.

New biofilter media are not always neutral. Their pH can be measured by mixing a small amount in water and measuring the pH of the water (Chapter 8). This mimics the process that will occur as the medium is put to use and shows the pH to which the microorganisms will be exposed. Some compost or peat will have pHs as low as 4 or 5 because of the organic acids produced during the natural decomposition of the organic matter. New activated carbon may have a pH as high as 10. The pH of a new medium should always be measured and adjusted to the design value.

The pH in a biofilter may also change during operation. Many of the contaminant biotransformations which occur in biofilters generate acids. The best approach for dealing with the acids may depend on whether these are final products or intermediates that will soon be degraded to non-acidic products.

For example, degradation of the alkane constituents of gasoline produces acetic acid as an intermediate (Dragun, 1988). Acetic acid is itself readily degradable and is usually consumed before concentrations rise enough to cause a significant pH decline. In some cases, however, a sudden spike of input contaminant, a change of temperature, or some other upset can cause acetate degradation to lag behind acetate production, so that acid accumulates and the pH falls. If the decline in pH further upsets the system, this may begin a self-reinforcing chain of events where the disruption becomes worse as more and more acid accumulates. Devinny et al. (1995) have described such a process in biofilters overloaded with ethanol. A similar phenomenon has been recognized for many years in the operation of anaerobic sludge digesters: when the "acid producers" get ahead of the "acid consumers", the pH may fall and cause the digester to fail.

Such episodes can be controlled by using a biofilter medium with a high buffer capacity. "Buffer capacity" is the term used to describe the ability of the medium to resist pH changes. Inorganic media may have low buffer capacities. To consider an extreme, a biofilter filled with glass beads would have none: if 0.9×10^{-6} moles of acid were generated per liter of water in the biofilter at pH 7, the hydrogen ion concentration would reach 1×10^{-6} moles per liter, and the pH would decline to 6. In a compost with moderate buffer capacity, some of the acid would react with the components of the medium, and the pH change would be smaller. At the other extreme, in a biofilter medium which includes calcium carbonate buffer, virtually all of the acid would react with the carbonate and the pH might change very little.

The buffer capacity of the medium can be defined as the amount of hydrogen ion which must be added in order to change the pH. In differential form:

$$\beta = \frac{dC_B}{dpH} \tag{4.3}$$

where β = buffer capacity; C_B = added acid or base.

Thus, when the buffer capacity is high, a large amount of added acid or base will have a small effect on pH. A high buffer capacity in a biofilter will make it resistant to pH changes cause by production of acid metabolites.

A related measure, the alkalinity, is defined as the amount of acid which must be added to reduce the pH to a chosen endpoint. It can be thought of as the "acid-neutralizing capacity" of the medium and can be determined by a straightforward laboratory titration (Chapter 8). It is commonly defined as the amount of acid necessary to reduce the pH to 4.5, but this may not be appropriate for a specific biofilter. If data indicate, for example, that treatment is inhibited when the pH falls below 5.5, then this endpoint should be used to determine the alkalinity of the medium for the case at hand. Alkalinity (pH 5.5) would indicate the amount of acid the biofilter could tolerate without inhibition.

Including buffering materials such as calcium carbonate in the medium can be an effective means for controlling occasional upsets. Acids which are produced are immediately neutralized, so the pH remains relatively constant, and time is available for the acid-consuming microorganisms to grow and consume the acidic compounds. As they are degraded, the buffer capacity may be restored so that the system can work indefinitely.

Degradation of chlorinated hydrocarbons produces hydrochloric acid as a final product. Treatment of hydrogen sulfide or organic sulfides produces sulfuric acid. Both of these products are stable end-products in an oxygen-rich biofilter, and both are highly soluble in water with very low Henry's Law constants. The biotransformations are thus very effective at removing the chlorine and sulfur from the air stream, but in both cases acids accumulate because there is no transformation which consumes the acids. Because the contaminants inevitably produce acid, and often the contaminant is present in significant concentrations, acid generation may be rapid. Inevitably, pH will fall to the point where the microbial ecosystem in inhibited.

In these cases, the addition of buffer to the medium may not solve the problem. Because the buffer is consumed steadily and rapidly, and there is no restorative phenomenon, it may be depleted quickly, requiring replacement of the medium. There may also be secondary problems. In one case, dissolution of the calcium carbonate generated small particles which contributed to clogging of the biofilter (Yang and Allen, 1994).

Fortunately, the alkalinity of the medium can be readily measured and compared to the amount of acid which will be generated by degradation of the pollutants in the waste stream. The designer can calculate whether it is possible to include enough buffer to last for the design lifetime of the medium. If the buffer will be consumed in 6 months, and the medium is expected to last for 4 years, another solution will be required.

Medium pH can also be controlled by adding base with the irrigation water. Because acids tend to accumulate at the inlet where the most biological activity occurs, the biofilter design with air in the up-flow mode and a lower soaker irrigation system is particularly well suited for this purpose. Again, the amount of base necessary can be readily calculated, and monitoring can show if too little or too much is being added; however, maintaining the uniformity of the pH in the medium may be difficult, because the irrigation water may not trickle through it uniformly. If a waste stream is expected to produce very large amounts of acids, it may be better to use a biotrickling filter. In these, the large flow of water provides efficient contact with all surfaces, and the pH of the recirculated flow can be easily and precisely controlled. In the case of sulfide treatment, biofilter operation may be more easily stabilized at low pH.

4.5 Nutrients

The microorganisms in biofilters consume contaminants for the energy and carbon they provide; however, they also need mineral nutrients: nitrogen, phosphorous, potassium, sulfur, calcium, magnesium, sodium, iron, and many others. Some species may require special compounds, such as vitamins, that they cannot synthesize for themselves. Successful biofilter operation requires that the needed nutrients be provided in the form and quantities that will support vigorous microbial activity.

Compost has the important advantage that the nutrients are present in the medium. Compost is made from plant tissues, sewage sludge, or other once-living tissues. As these undergo degradation, just those elements and compounds which are needed for life are released. The nutrient needs of microorganisms are similar to those of plants, so compounds and elements are released during degradation in the approximate proportions appropriate for cell growth. However, it is possible that the rate of degradation, and therefore the rate at which soluble nutrients are generated, can be too slow. Gribbins and Loehr (1998) found that nutrient release rates were limiting in a compost-perlite biofilter treating a high load of toluene.

Inorganic media such as rock, activated carbon, plastic shapes, or polyurethane foam do not contain an appropriate supply of nutrients. These must

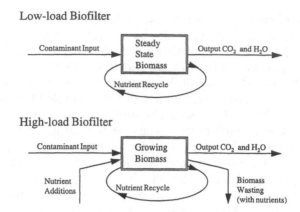

Figure 4.5   Comparison of biofilter operating regimes.

be added when the biofilter is put into operation, usually in the form of commercial fertilizers added to the irrigation water. Typically nitrogen, phosphorous, and potassium are added.

When the biofilter medium is prepared, it is also possible to add slow-release fertilizers. Many kinds have been developed for agricultural use with the intent of reducing the necessary frequency of fertilizer addition. In biofilters, the release replenishes nutrients as they are lost to leaching or biotransformation. Monitoring is still necessary, however, to ensure that replacement rates are as high as loss rates.

Biofilter operators must maintain the continuing availability of nutrients as operation proceeds. Ideally, a biofilter has a stationary water phase and a steady-state microbial ecosystem, so that it might be expected that the nutrient content would be maintained and continually recycled. Degradation of the biomass releases the nutrients in soluble form, where they can be taken up again by growing cells (Figure 4.5). However, biofilters sometimes produce leachate, either intentionally or inadvertently, and this will carry dissolved nutrients out of the biofilter. Further, if the biomass grows rapidly, it may tie up all the nutrients, reducing soluble concentrations and preventing further growth. Anaerobic activity, occurring in deep pores or at the base of the biofilm, can convert available forms of nitrogen to nitrogen gas that will be lost to the air flow. Ammonia is produced by degradation of the medium or biomass, and can also be lost as the gas.

Gribbins and Loehr (1998) determined that the highest treatment rates in a compost-perlite biofilter were partially limited by soluble nitrogen availability unless the concentration was 1000 mg kg^{-1}, and that the nitrogen-to-carbon ratio should be at least 1 to 100. These are much higher than previously reported and may indicate that nutrients are unrecognized controlling factors at high loads. They further cautioned that leachate production should be limited to reduce nitrogen losses. Because nitrogen uptake varies with contaminant degradation rate, nitrogen availability may be very different at different levels within the biofilter, so thorough sampling may be necessary to assess nutrient limitation.

For these reasons, it is prudent to replenish nutrient concentrations from time to time. Ideally, the medium could be analyzed, and appropriate amounts of nutrient added as needed. More typically, operators simply add fertilizers on a regular schedule.

It is possible to imagine much more obscure relationships. The one species of microorganism which is best for a particular contaminant, for example, might have an absolute requirement for vitamin B_{12}. Understanding of these patterns awaits more research.

4.6   Contaminant load and surface load

The mass of contaminant entering the biofilter per unit time and per unit volume is the contaminant load. It has major effects on biofilter operation, and it is important to interpret the results of experiments and operational experience in light of the load being treated. Biofilters employed for odor control purposes, where concentrations are typically a few parts per billion, may treat less than 1 g m^{-3} h^{-1} of contaminant. At the other extreme, biofilters treating industrial effluents or soil vapor extraction off-gases may see concentrations in the thousands of parts per million and loads above 100 g m^{-3} h^{-1}. Removal efficiency may be excellent when loads are low, but poor when loads are high. Acidification of the medium, and ultimately system failure, is far more likely when the loads are high, and the acid-generating steps in the degradation may run ahead of the acid-consuming ones.

Biofilters which are operating at low loads may reach approximate steady state with respect to biomass and nutrients (Figure 4.5). The microbial ecosystem is starved for substrate, and predator species will consume biomass as it is produced. Biofilters treating very high loads will grow biomass rapidly if nutrients are available. They will require a means for removing the biomass or otherwise preventing excessive pressure drop across the bed. Backwashing with water has been done for a biotrickling filter (Sorial et al., 1997). Because the wasted biomass contains a substantial amount of nutrients, more nutrient must be continuously supplied. Of course, some biofilters may be operated in the middle range, with occasional cleaning, medium replacement, and nutrient resupply required.

A given contaminant load may be delivered in a small amount of slow-moving air or in a large amount of fast-moving air, i.e., it may be a high concentration in a low surface load or a low concentration in a high surface load. Under many conditions, the higher concentration will be treated more effectively. It will produce a higher concentration of contaminant in the biofilm, speeding biodegradation if the biological kinetics are higher than zero order. In a case of diffusion limitation, the higher air concentration will drive the contaminant into the biofilm more rapidly.

Variations in load are common in real applications. Industrial processes may be operated at greater or lesser rates as the day progresses, and many will be shut down at night and on weekends. Biological systems operate best, however, on steady loads (Chapter 5). Some load equalization occurs as a result of the adsorptivity of the biofilter medium. Deshusses (1997) reported that a step increase in the load of methyl ethyl ketone (MEK) in a laboratory biofilter resulted in increases in carbon dioxide release only after a few hours' delay. He suggested that the contaminant had been adsorbed immediately, but was degraded over the course of a few hours. Loy et al. (1997) noted that CO_2 production in a biofilter did not return to maintenance levels until 42 minutes after feed concentrations of toluene were reduced to zero.

More elaborate efforts at load equalization may be also be possible. A short detention time, granular activated carbon adsorber can be placed in the influent stream (Swanson and Loehr, 1997). It will adsorb contaminant when peak concentrations appear, then release them when the influent is cleaner. Including carbon in the medium mix will have the same effect (Weber and Hartmans, 1995; Leson and Smith, 1997). Influent may be split and fed to different levels in the biofilter bed, so that a peak load is spread over more of the medium.

4.7   Oxygen limitation

In high-performance biofilters, oxygen limitation may occur in the biofilm. Note that the term "oxygen limitation" does not necessarily imply that oxygen is completely depleted in the biofilm, but refers to the general situation where the rate of biodegradation is affected by the concentration of oxygen (see Equation 4.4 below). At first, the existence of oxygen limitation in an air biofilter might sound contradictory because air contains 21% oxygen. However, the reason for oxygen limitation is that the oxygen gas-liquid partition coefficient is 33.5, meaning that most of the oxygen is in the gas phase rather than dissolved. For example, at 25°C, the dissolved oxygen concentration in equilibrium with air is about 8.1 mg L^{-1}, or 0.253 mMol.

In a first approximation, the question of oxygen depletion in the biofilm is relatively simple to address using diffusion and partition coefficients and the stoichiometry of oxidation. The oxygen diffusion coefficient in water is about 2.1 10^{-9} m^2 s^{-1}, while common VOC diffusion coefficients are on the order of 0.8 to 1.3 10^{-9} m^2 s^{-1} (Cussler, 1997). This is not greatly different, so in the following reasoning oxygen and VOC diffusion coefficients are assumed identical. Consequently, if the stoichiometric amount of oxygen necessary to degrade the treated VOC (using the VOC gas/liquid interfacial concentration) is larger than about 0.25 mMol (the interfacial concentration of oxygen), chances are that oxygen will be exhausted in the biofilm before the VOC is treated. In other words, this suggests that the effective biofilm thickness, as defined by Williamson and McCarty (1976) and discussed in Chapter 6, will be determined by oxygen rather than VOC availability.

Table 4.1   Estimated Threshold Concentrations for Oxygen Limitation and Maximum Elimination Capacity for the Treatment of Various VOCs

Compound	Dimensionless Henry's coefficient	Amount of oxygen needed for complete oxidation	Lowest concentration of contaminant in the air to induce oxygen depletion in biofilm[a] $(g\ m^{-3})$	Maximum VOC elimination capacity, assuming an oxygen transfer rate of $200\ g\ m^{-3}\ h^{-1}$
Ethanol	0.000257	3	0.0009	88
Ethyl acetate	0.0055	5	0.0245	110
Toluene	0.275	9	0.7112	64
Hexane	74.13	7.5	195	65

[a] Calculated assuming complete aerobic oxidation; oxygen and VOC diffusion coefficient to be identical; a temperature of 20 to 25°C.

Oxygen deprivation is undesirable because it can lead to partially oxidized by-products, such as carboxylic acids (Devinny and Hodge, 1995) or aldehydes which can cause nuisance odors or system upset. Some values for the theoretical minimum VOC concentration that will cause oxygen limitation are reported in Table 4.1. The values show that treatment of hydrophilic compounds, due to a more favorable partition in water, is most probably limited by oxygen at concentrations normally prevailing in biofilters, whereas treatment of hydrophobic compounds will probably not be subject to oxygen limitation. This is not commonly observed in biofiltration experiments, or at least is not commonly identified as such. This is probably because a number of other complex factors come into play, and because investigators have so far found other plausible explanations for the phenomena they have observed. Even so, Kok (1991) calculated the maximum elimination capacity of biofilters for various VOCs based on a hypothetical oxygen transfer rate of 200 g m⁻³ h⁻¹ and found maximum elimination capacities to be well in the range of commonly observed values (Table 4.1). This suggests that oxygen limitation might be more common than investigators suppose.

One problem with this approach is that it is based on an idealized biofilm geometry (flat) and properties (homogenous) and is extrapolated to a system (the biofilter) which is significantly more complex. For example, the work of Møller et al. (1996), showing highly non-homogenous biofilm structures suggests that the porous, three-dimensional, channel-like structure of the biofilm plays a major role in enhancing the transfer of both the pollutant and the oxygen. In a previous study with submerged biofilms, de Beer et al. (1996) demonstrated that the supply of oxygen through such voids and channels was roughly 50% of the total oxygen transfer. Consequently, the real VOC concentration thresholds for oxygen limitation might be very different than those reported in Table 4.1. Even so, the values of Table 4.1 also suggest that if oxygen limitation is expected, it might be beneficial to dilute the air prior to treatment. This goes against the common practice and requires further experimental verification. A more in-depth analysis of the form of various biodegradation rate equations shows that dilution might actually have the opposite effect, if kinetics are first order in pollutant concentration.

Oxygen limitation may influence biodegradation rates even if oxygen is not completely depleted in the biofilm. This case is usually referred to as a double limitation and biodegradation rates can be modeled using a double Michaelis-Menten type kinetic:

$$k = k_{max} \frac{C_{L,j}}{K_{M,j} + C_{L,j}} \cdot \frac{C_{L,O}}{K_{M,O} + C_{L,O}} \qquad (4.4)$$

In this case, both the substrate-VOC and the oxygen play a role in the determination of the biodegradation rate. Determination of the parameters in Equation 4.4 is relatively easy in a shake flask or a bioreactor with suspended cultures. However, experimental determination of those parameters in biofilters is nearly impossible because of the inability to measure or control conditions in the microenvironments within the biofilm. Clearly, the absence of simple experimental methods to investigate the effects of oxygen concentrations on pollutant removal kinetics poses a challenge. Identification of oxygen limitation in full-scale biofilters and meaningful measures to remediate the limitation remain objects of further study.

Actual oxygen profiles have been experimentally measured only in a limited number of cases (Mirpuri et al., 1997). They showed that oxygen could indeed become limiting. Because of the difficulty of direct micrometric measurement of dissolved oxygen, other efforts were directed towards indirect demonstration of oxygen limitation. This was usually done in the laboratory by varying the oxygen content of the treated air, while monitoring pollutant removal. In some instances oxygen limitation could be demonstrated (Cox et al., 1997), in others, increasing the oxygen content in the air did not influence the elimination of the treated pollutants (Deshusses et al., 1996).

Another type of oxygen limitation may occur in biofilters. This is when part of the biofilter bed has a much lower air permeability than the rest of the bed, as a result of bed compaction, improper air distribution, or medium loading. "Pockets" where only limited air exchange takes place are formed. Over time, unwanted anaerobic conditions will develop in these pockets. For the biofilter operator, it is rather difficult to determine whether a fraction of the bed is anaerobic. Emission of metabolites or malodorous compounds and medium acidification are signs of anaerobic activity. If dead zones constitute an important fraction of the bed, reduced performance will be observed. The ultimate proof that dead zones are present is given by residence time distribution testing, either by pulsing a tracer gas into the inlet or by performing smoke tests in open beds (Chapter 8). Unfortunately, there is no easy remedy to the existence of anaerobic pockets. If nuisance odors are released or emission guidelines are violated, the biofilter medium must be unloaded and carefully repacked. In many instances, repacking the biofilter will provide an opportunity to inspect and maintain the biofilter structure, as well as possibly replacing the medium.

It is difficult to define situations specifically in which oxygen limitation will occur; however, oxygen limitation is more likely to occur in the case of high concentrations of easily degradable hydrophilic compounds and in systems where thick biofilms exist. From an operator standpoint, there is no proven way to remediate to oxygen limitation. Adding oxygen to the inlet air is unlikely to be economically feasible, and diluting the inlet air may result in unexpected effects but may be worth bench-scale investigation and treatment cost evaluation. The most rational option is to repack the biofilter so that the interfacial area and oxygen transfer are maximized.

4.8   Air flow direction

In most applications of enclosed biofilters, downward air flow has proven superior to up-flow. Even so, there are a number of success stories of biofilters operating in an up-flow mode, meaning that down-flow direction is not a *sine qua non* design condition. The advantage of down-flow is that it improves moisture control. If drying of the biofilter medium should occur, it will generally start from the inlet side of the biofilter as the result of either unsaturated inlet air or production of metabolic heat concentrated at the inlet side. In the case of downward flow, moisture can be efficiently controlled by additional water supply provided as a spray on top of the bed, where it is most needed. In the up-flow mode, drying occurs preferentially at the bottom where it is difficult to provide additional moisture. Another reason to prefer the down-flow mode is that it allows better drainage, particularly at the bottom of the bed. In the case of upward air flow, great care should be given to drainage and to air distribution systems (Chapter 7).

There are, however, several cases in which up-flow is beneficial. During treatment of reduced sulfur compounds (e.g., odors from a treatment plant), sulfuric acid is generated by sulfide-oxidizing organisms. This causes the pH to decrease, particularly near the air inlet where the biological activity is concentrated. Similar phenomena are observed for the treatment of chlorinated compounds where chloride formation will cause the pH to drop. Excessive pH drop is detrimental to pollutant removal. In such a case, upward flow might be preferred, enabling trickling of water or a pH buffer from the top of the biofilter or, even better, from a lower irrigation system. The acidic end products, sulfate, chloride, etc., are then easily washed out without leaching through the entire bed, as would occur if sulfate were concentrated at the top of the bed in down-flow systems. Further, the usual

problems associated with moisture control in up-flow biofilters can be reduced by the installation of a lower irrigation system.

A few alternating air flow direction systems have been proposed. Apart
from some technical-scale biofilters installed in Germany (Sabo et al., 1996),
these developments remain essentially laboratory or pilot scale. Switching
the air direction in a fully enclosed container biofilter has been proposed
(Sabo et al., 1996). The purpose is again to control moisture content better, but
it has also been used by others to control the growth of biomass by starving
a portion of the biofilter (Kinney et al. 1996). In the former application, the air
normally flows upward. The flow is only reversed during the moistening
phase, when the overbed sprinklers are turned on. A comparative study of
conventional systems and directionally switching systems is yet to be performed to assess the true benefits of the technology objectively. Other bed
geometries have also been proposed (Bodker and Rydin, 1996; Sabo et al.
1996). Besides the Rotor-Biofilter (Sabo et al. 1996), field application is miss-
ing, and demonstration of the benefits of such systems is needed.

4.9   Dust and grease

Biofilters are designed on the assumption that the material they receive can
be completely converted to gaseous or soluble products which are carried out
of the biofilter. They are not designed to handle dust. In the short term, they
will efficiently collect dust, because the air flow is divided into fine tortuous
streams with a large surface-to-volume ratio. As the air passes through,
opportunities for dust particles to contact the walls of the pores are many.
The medium is wet, and a particle that contacts it will likely stick. Thus, dust
will be removed from the air, but it will simply accumulate in the medium.
If large amounts of dust are present, the biofilter will soon clog, and replacement of the medium will be required. Unfortunately, there have been problems with a few large-scale biofilters treating dusty air from the pressboard
industry, because the accumulation was not anticipated. Shortly after startup,
it was necessary to install wet scrubbers to pretreat the air (Allen and Van Til,
1995). Togna et al. (1997) found clogging in a biofilter designed for treated
wood products waste and decided it had resulted from the accumulation of
resinous lignin degradation products and benzoic acid. These were carried
into the biofilter either as vapors which condensed or as aerosols.

Particulates or aerosols may be biodegradable, and it is conceivable that
they could be collected and biodegraded. This might be important, for example, in biofilters designed to treat cooking odors. Cooking discharges
contain substantial amounts of grease aerosol. While this is certainly biodegradable, the amounts may be too large. The mass of contaminant per unit
volume of air can be much higher for an aerosol than for a ppm-level vapor.
Accumulations of grease have clogged biofilters used for treatment of food
preparation exhaust. Because grease can be efficiently collected in particle
collectors, it is appropriate to have these installed upstream of the biofilter.

In any case, the design engineer must provide an adequate pretreatment system to remove grease and dust. This may make biofilters uneconomical for some applications. Conceivably, a biofilter could treat a mixed waste of biodegradable vapors and nonbiodegradable dust, using occasional backwashing to remove the dust; however, there have been no reports of such systems.

4.10 Extreme biofilters

For the most part, biofilter designers and researchers have made systems that operate under the most benign possible conditions. To maximize treatment rates, every effort is made to keep the pH near neutral, the temperature moderate, and water and oxygen abundant. This approach is generally successful. Certainly these conditions support a wide variety of microorganisms with high metabolic rates. Aerobic metabolism releases a large amount of energy per mole of material processed and so is generally carried out rapidly by microorganisms seeking to maximize their growth and ecological success.

Nature, however, teaches us that microorganisms can be successful under a very wide variety of conditions. They are found in acidic hot springs at temperatures over 90°C and at deep sea thermal vents at temperatures over 100°C. Sulfide-oxidizing microbes thrive where the pH is 1, and some fungi can grow in environments where the soil water suction is as high as –681 bar (Stolp, 1988). Many species carry on life and transform organic chemicals in the absence of oxygen. It is reasonable to suppose that some of these microorganisms, under the appropriate biofilter operating conditions, could be useful for treatment purposes.

4.10.1 Low pH biofilters for sulfide oxidation

Hydrogen sulfide is a highly odorous and toxic gas produced in wastewater collection and treatment facilities and in other industrial systems. Its release to the atmosphere brings odor complaints, and if it collects in confined spaces, it can be deadly. Under aerobic conditions, there are many species of organisms which can oxidize it to produce sulfuric acid. Acid formation on the crowns of wastewater pipes can cause rapid corrosion, and pipes, buildings, and other appurtenances in treatment facilities have sometimes been seriously damaged.

These problems often appear in biofiltration of wastewater off-gases. Conversion of the hydrogen sulfide to sulfuric acid removes it from the air, solving the odor problem. The solubility of hydrogen sulfide is high, and conversion rates are rapid, so biofiltration is an effective treatment process. But, the acid produced causes the pH of the biofilter to fall, and some investigators have seen substantial reductions in treatment success (Yang and Allen, 1994; Furusawa et al., 1984). This is often countered by the addition of buffering materials to the medium, or by the addition of base with the irrigation water.

An alternative is to allow the biofilter to operate at low pH. A series of species of genus *Thiobacillus* is capable of oxidizing hydrogen sulfide in environments of successively lower pH (Islander et al., 1991). Below pH 3, systems are often dominated by *Thiobacillus thiooxidans*, which oxidizes sulfide rapidly. *T. thiooxidans* is not inhibited until the pH falls below 1.

A sulfide biofilter may be operated at low pH, with several advantages. *Thiobacillus* does not produce exopolysaccharides, perhaps because oxidization of sulfide produces less energy per mole of substrate than oxidation of organic materials. The microorganisms are autotrophic, making their organic matter through the energy-consuming process of fixing carbon dioxide. This means that a pH 1 biofilter is less susceptible to clogging by an overgrown biofilm. Several transformations are necessary to convert sulfide to sulfuric acid, and at low pH some of these will occur by rapid chemical processes (Islander et al., 1991). Finally, *T. thiooxidans* is accompanied by acidophyllic heterotrophs. These are symbionts that consume low-molecular-weight fatty acids which are waste products of the *Thiobacillus*. Indeed, if the heterotrophs are not present, *Thiobacillus* is soon self-inhibited by the accumulation of these compounds. But, the heterotrophs may also be able to consume some of the organic compounds in the air stream, so that a sulfide biofilter can treat more than just sulfide. One study has shown that a low-pH biofilter is effective for simultaneous treatment of low concentrations of reduced sulfur compounds and VOCs from a wastewater treatment facility (Webster et al., 1996, 1997).

Operating the biofilter at low pH may have another advantage. The easiest way to control the pH in a biofilter where acid is being produced is to wash it away with an excess of irrigation water, producing some leachate. However, at pH 7, where the acid concentration is low, substantial amounts of water are necessary to remove the acid produced by the typical sulfide load. At pH 1, the acid concentration is a million times higher, so that a million times less water is necessary to carry the acid away. Water washing for pH control is not practical at pH 7, but is easy at pH 1.

4.10.2 *Anaerobic conditions in biofilters for chlorinated hydrocarbons*

There are many biochemical transformations that occur only under anaerobic conditions and are useful in waste treatment. Anaerobic digestion of sewage sludge is common, and anaerobic treatment of concentrated industrial wastes is commonly done in fixed-film reactors. Anaerobic biotransformations do not use diatomic oxygen to convert organic materials to carbon dioxide and water and so produce substantially less energy per mole of compound transformed. This slows the workings of the microbial ecosystem, and anaerobic waste treatment is generally slower than aerobic treatment. If both aerobic and anaerobic treatment are possible, aerobic conditions will be chosen where rapid treatment is desired, and anaerobic conditions where low sludge production is preferred.

Anaerobic treatment is common because there are many compounds which can only be attacked by microorganisms under anaerobic conditions. Anaerobic treatment of sewage sludge consumes many of the compounds not removed in the earlier aerobic treatment. In particular, some chlorinated hydrocarbons, such as tetrachloroethylene (also called perchloroethylene or PCE), can only be degraded anaerobically.

While the gas phase in biofilters treating air streams will inevitably be aerobic, portions of the biomass may not be (Section 4.7). If biodegradation within the biofilm is consuming oxygen, it is quite possible for the gas to be exhausted in the deeper portions of the biofilm. Because the biofilm is highly irregular, occupying the corners and pockets between the packing particles, some portions will be far more distant from the air phase than others. These may easily become anaerobic. Compounds can be transferred from the air and diffuse through the biofilm to the anaerobic zones where they will be biodegraded. Products will diffuse outwards. As in all biotransformations, many steps may be involved. The compound might undergo some aerobic reactions at the surface of the biofilm, with the products diffusing inwards to be further degraded anaerobically. Products from this reaction might undergo further aerobic transformation as they diffuse outwards. Because anaerobic transformations tend to be slower and because the process may be limited by diffusion rates, it can be expected that anaerobic degradation within an aerobic biofilter will be slow.

A similar process has been postulated as the explanation for degradation of PCE that was observed in laboratory-simulated aerobic groundwater. Enzian et al. (1994) suggested that degradation of the chlorinated hydrocarbon was occurring in anaerobic zones within the soil.

PCE is found in off-gases from wastewater treatment plants and in air discharges from air strippers used for groundwater treatment. Because it is present in low concentrations in a large volume of air, biofiltration seems a desirable solution; however, oxygen is inevitably present in abundance in these air flows and in almost all biofiltration applications. Biofiltration at first seems an unworkable technology for PCE. Nevertheless, biofiltration has been successful at removing low concentrations of PCE and may be applicable to other compounds for which anaerobic degradation is possible (Devinny et al., 1995).

Treatment of PCE in off-gases from a wastewater treatment plant has been reported (Webster et al., 1996, 1997; Devinny et al., 1995). Low incoming concentrations (a few ppb to hundreds of ppb) of PCE and trichloroethylene (TCE) were successfully removed in an activated-carbon-filled biofilter at moderate pH. An initial period of very good treatment likely represented simple adsorption but was followed by an extended period of biological removal of about 61% of the PCE and about 48% of the TCE on carbon biofilters.

Anaerobic degradation of chlorinated hydrocarbons generally proceeds by reductive dechlorination (Norris et al., 1994). An electron donor, generally a readily degradable organic compound, is also necessary. In the case of the wastewater off-gases, the electron donors were presumably some of the organic contaminants which were also present in the air. In the laboratory, Devinny et al. (1997) used a carbon-based biofilter supplemented with lactose to achieve 70 to 90% removal of 10 to 40 ppm of PCE in a detention time of 3.7 minutes.

An entirely anaerobic biofilter could have applications for gas streams which do not contain oxygen. Landfill gas is dominantly methane and carbon dioxide but may contain trace amounts of chlorinated hydrocarbons. These interfere with utilization of the gas for energy because the chlorine causes rapid corrosion of metals in machinery such as turbines. Devinny et al. (1997) have shown that biofiltration can be used to remove the chlorinated hydrocarbons in a system which is entirely anaerobic.

Anaerobic biofiltration for treatment of carbon tetrachloride was demonstrated by Lee et al. (1997). They added methanol vapors to serve as an electron donor and successfully removed carbon tetrachloride from a helium gas stream. The necessary detention time, however, was very long, and removal efficiencies were low and variable, generally below 50%.

4.10.3 Biofilters using cometabolism

Some compounds that are generally recalcitrant to biodegradation can nevertheless be degraded cometabolically. In cometabolism, microorganisms which are utilizing a growth substrate as their energy source have been found to fortuitously degrade unrelated compounds, probably because they have a similar shape which fits the active site of the enzyme. Typically the organism gains no energy or other benefits from degrading the cosubstrate. Cells growing on methane, toluene, or phenol, for example, can degrade TCE.

The discovery of cometabolism immediately suggested possibilities for treatment of recalcitrant compounds. If the growth substrate is supplied to keep a culture active, the microorganism will degrade the contaminant, as a cosubstrate, at the same time.

This approach has been used in biofiltration of air in a few cases. Speitel and MacLay (1993) used a laboratory-scale biofilter supplemented with methane to treat TCE. While the approach successfully degraded TCE, some difficulties were encountered. When the methane concentrations were high, the microorganisms grew vigorously, but the methane competed so effectively for the key enzyme that degradation rates for TCE were low. When the methane concentration was reduced, TCE degradation rates initially rose. But, without methane to support the microorganisms and to induce production of the enzyme, its concentration fell. Within a few hours, TCE degradation also declined. The investigators suggested, however, that it would be possible to operate the biofilter on an alternating cycle, with and without methane. Two biofilters could be run in parallel with alternate timing and the TCE stream passed through the one that was not receiving methane at the time.

Hecht et al. (1995) designed a laboratory-scale bubble column biofilter to treat TCE, using phenol as the growth substrate. They achieved 30 to 80% removal of the solvent vapors. The phenol was fed slowly into the water in the column, so that it was operated as a chemostat. It was possible to maintain a working culture with phenol concentrations half those of the aqueous TCE, so competition for the enzyme was not a serious problem. While this system was not a biofilter, it suggests that a biofilter could be operated using similar metabolic mechanisms.

4.10.4 NO$_x$ biofilters

Davidova et al. (1997) demonstrated aerobic biofiltration of nitric oxide using microorganisms of the genus *Nitrobacter*. These bacteria are capable of oxidizing nitric oxide to nitrite and nitrate and were obtained in inocula from a nitrifying wastewater treatment plant. Under aerobic conditions, this transformation is thermodynamically favored and can provide the energy needed for microbial growth. The biofilter achieved 70% removal of NO$_x$ at concentrations of 80 ppmv with an empty bed contact time of 12 minutes. The rather long EBRT required is consistent with the generally lower metabolic rates of microorganisms utilizing low-energy mineral substrates.

Because neither oxidation nor reduction of nitric oxides provides carbon, the organisms must obtain it from elsewhere for their growth. In aerobic systems, the microorganisms can obtain their carbon by fixing CO$_2$. Davidova et al. (1997) recognized that this is an energy-consuming step and that it may have slowed the activity of the *Nitrobacter*. They increased the concentration of CO$_2$ in the gas stream in order to reduce the amount of energy the microorganism would have to spend on fixation; however, this also reduced the pH of the water, and the net effect was a decline in nitric oxide removal.

Apel et al. (1995) investigated anaerobic removal of nitrogen oxides from combustion gases using denitrifying bacteria which convert the oxides to nitrogen gas. Under anaerobic conditions, this is again thermodynamically favorable, and removal of 90% of the NO was possible. However, in a gas stream with a modest amount of oxygen present (5%), removals dropped to 39%. Apel et al. also added molasses to provide carbon. Its oxidation by other organisms may have helped maintain anaerobic conditions within the biofilm when oxygen was present in the gas stream.

Du Plessis et al. (1996) also investigated anaerobic mechanisms for nitric oxide removal, but they did so within a dominantly aerobic biofilter. They utilized a thick biofilm which created anaerobic underlayers, allowing denitrification to nitrogen gas; 75% removal was obtained at an EBRT of 6 minutes.

4.10.5 High-temperature biofilters

Most chemical reactions, including biochemical transformations, occur more rapidly at high temperatures. Biological treatment processes using thermophilic organisms take advantage of this to create more rapid and therefore more economical treatment processes. Anaerobic digestion of sewage sludge is often done at temperatures greater than 60°C. While heating a biofilter is unlikely to be practical, biofiltration can be done at elevated temperatures if the incoming air is warm. Indeed, cooling the air to ambient temperatures would be expensive. This is the case for many industrial processes.

At least one biofilter has been designed, inoculated, and operated specifically for thermophilic operation. To treat hot gases containing ethanol, van Groenestijn et al. (1995) used a biofilter filled with porous ceramic particles. It was inoculated with a culture taken from the thermophilic stage of a composting process and operated with water in the pores of the ceramic but not on the surface of the particles. Fungi grew vigorously, with some mycelia extending into the ceramic to obtain water and nutrients and others extending into the air to collect the ethanol. The biofilter achieved excellent performance, removing 80 g m^{-3} h^{-1} of ethanol. The fungi were truly thermophilic, and treatment dropped off sharply at temperatures below 60°C. Treatment was restored within 6 hours after a shutdown of 48 hours. This is a slower recovery than mesophilic biofilters, presumably because of the highly specialized nature of the fungi, but may be acceptable in many applications.

Elevated temperatures, however, also reduce the solubility of most gases, and it is possible that this would reduce contaminant concentrations in the biofilm sufficiently to cancel some of the benefits of higher degradation kinetics. Decomposition of degradable support media may also occur more rapidly. Deshusses et al. (1997) operated a biofilter treating ethyl acetate at 45 to 50°C. It achieved removal efficiencies greater than 100 g m^{-3} h^{-1}, comparable to performance at 30 to 37°C.

4.10.6 Low-water-content biofilters

Observations of filamentous fungi in biofilters are common. It is likely that all biofilters support some fungi among the complex suite of organisms which carries on the treatment process. While they are generally less active in terms of mass of material degraded per unit volume of organisms per unit time, they tend to display greater metabolic diversity, attacking many compounds which are beyond the capabilities of bacteria. They are also more tolerant of drying. Biofilters are usually operated with water activities near 1.0 to provide a benign environment for bacteria, but some fungi can remain active where water is much less available. *Xeromyces bisporus* can grow in soils with water activities as low as 0.61 (Stolp, 1988). van Groenestijn et al. (1995) have reported a successful bench-scale biofilter for treatment of toluene, ethylbenzene, and *o*-xylene, dominated by fungi and operated at low water contents.

Fungi can withstand drier conditions than bacteria partly because of their metabolic adaptations, but they also differ because they can develop macroscopic structures. The hyphae, or filaments, grown by fungi may serve almost as "roots", reaching into the crevices and pores of the support medium to obtain water from the wetter environments, while other parts of the organism are at the surface collecting contaminant for degradation. It is intriguing to ask whether the fungi that grow filaments that stand out from the surface of the medium can draw substrate directly from the air. A patch of fungal "fuzz" has a very high surface-to-volume ratio for collecting vapor-phase contaminants.

While properly controlled growth of filamentous fungi may be useful, unplanned growth may block the pores in biofilters and cement the particles of medium together. Utilization of fungal biofilters may require special design, such as specification of a medium with larger particles.

chapter five

Microbial ecology of biofiltration

5.1 Introduction

Microbiological activity transforms pollutants to harmless products in biofilters. The species which are present, their population densities, the metabolic transformations they are catalyzing, and their interactions with their environment and each other are fundamental to biofilter operation. The study of a diverse collection of organisms interacting with each other and their environment is ecology; the microorganisms in a biofilter can only be understood by considering them as part of an ecosystem. Our knowledge of the working of ecosystems is at best fragmentary, and our understanding of microbial ecosystems is weakest of all. Even so, an ecosystem approach is the best guide for efforts to develop better microbial communities for pollutant treatment.

5.2 Microbial species in biofilters

5.2.1 Selection and proliferation

Biofilters are inevitably biologically open systems. The very large amounts of air which pass through them carry aerosols and dust, and these in turn carry the cells, spores, and cysts of a tremendous variety of microorganisms. When compost is used as a biofilter medium, it brings with it an initial inoculum, including thousands of species. As biofiltration proceeds, these species will thrive or fail according to their abilities to find a place in the biofilter ecosystem. Even in a biofilter treating a single contaminant, there will be many ecological niches to occupy. Complex compounds may require many metabolic steps in the transformation from their original form to carbon dioxide and water, and different species may specialize in different parts of the process.

Species consuming the same substrate, whether it is the original contaminant or some metabolite, will compete fiercely. The less capable species may die out. But, there may also be specialization on the basis of biofilter microenvironment: some microorganisms might do well at the surface of the biofilm, while others succeed deep within it, and yet others swim in the water film outside of it. One species could do well in deep pores where water is more abundant, while another succeeds on the convex surfaces of the support medium, where the thinner water film excludes predators. If the biofilter removes more than one pollutant from the air, the number of degraders will be multiplied accordingly.

Bacteria and fungi are certainly the two dominant microorganisms groups in biofilters. Most biofilms will contain substantial numbers of both, but their relative abundances can vary widely. Bacteria have the advantage of rapid substrate uptake and growth. Under favorable conditions, they will dominate, although fungi will be present. Fungi generally grow more slowly, and their larger size gives them a smaller surface-to-volume ratio for substrate uptake. But, they are often capable of degrading a greater variety of contaminants and can withstand harsher conditions. van Groenestijn et al. (1995) showed that they could thrive where the pore water pH was 2.5, where the air in the biofilter was relatively dry, and even at temperatures between 60 and 71°C. de Castro et al. (1997) saw colonies of various kinds in biofilters using various inocula, and patches of fungus grew and waned as treatment proceeded. However, it is also important to note that fungi can produce tangled erect filaments which may block the air flow, and they may cement the medium particle together. Heavy growth of erect forms may increase head loss and interfere with medium tilling or replacement.

Predators are common in biofilters. In the microbial ecosystem, the contaminants become the food source which supports a variety of degrading microorganisms. As their populations grow, they will attract protozoa which consume them whole, bacteria which parasitize them, and viruses which kill them and release more viruses. Each degrader species may be food for several predators, and every predator may be victim to predators higher in the food chain.

Protozoa have been found in large numbers in biofilters. de Castro et al. (1997) saw numerous protozoa in bench-scale biofilters, ranging in size from 5 to 50 μm. Larger cells were seen early in the acclimation process, while smaller species were seen as the systems moved to steady state.

Ecologists have proposed that most ecosystems developing in a new habitat pass through a succession of structures before they reach a climax community in which species populations roughly stabilize. There are indications that the microbial ecosystem in a biofilter may not reach a steady state until long after it has reached high rates of biodegradation. Webster et al. (1996) characterized a biofilter microbial ecosystem using phospholipid fatty acid analysis and found that its characteristics became approximately constant only after hundreds of days.

5.2.2 Inoculation of biofilters

The open nature of biofilters limits the control designers and operators might wish to have. Many investigators have suggested using a single ideal species, known to vigorously degrade the compound of interest, as inoculum for a biofilter. Such an approach may be successful, but only if the species is also "ecologically viable". The species must be able to succeed under the environmental conditions of the biofilter, growing at pH and temperature prevailing in the reactor and withstanding the inevitable variations. It must be able to survive competitive organisms, avoid predators, and grow fast enough to make up its losses to both. Swanson and Loehr (1997) have noted reports of inoculation which had no effect on compost biofilters, and suggested that the inoculum species could not compete. The ability to degrade the pollutant is a necessary trait, and rapid utilization of the pollutant will give a species a competitive advantage, but it does not guarantee success. Indeed, some investigators have suggested that microorganisms grown in the lab and released for the purpose of bioremediation of soils may fail because they have become "lab adjusted". In as few as 25 generations, the species can develop traits which are ideal for the petri dish environment and lose those which provide fitness in the wild. Bohn (1996) was particularly blunt in saying, "Expecting an introduced laboratory or bioengineered microbial culture to flourish in a field conditions is naive. The cultures are originally derived from soils and fed special food sources in the laboratory. Food sources and predator relations are much different in the field. ... The new microbes are just another food source for the native population."

While there are many reasons why inoculation with a single chosen species may fail, it is unlikely to do any harm. The choice and preparation of a proper inoculum remains an important matter for future research.

An advantage of using compost for the biofilter medium is that it brings with it a well-developed community of microorganisms. The organisms which have been doing the composting are abundant and well adapted to their environment. The microbial consortia present can degrade the multitude of compounds found in leaves and grass. When the compost is placed in a biofilter and the air is passed through, the contaminant will become a dominant substrate. Those microorganisms which can degrade it will grow rapidly and become abundant. However, even compost may benefit from inoculation in some cases. Wright et al. (1997) found that acclimation in compost biofilters treating gasoline vapors was much more rapid when they were inoculated with a culture which had been grown on gasoline. Leson and Smith (1997) came to the same conclusion, suggesting that inoculation will speed acclimation but will not affect ultimate removal efficiencies. The delayed acclimation of the species found on the compost may result from compounds in gasoline which are uncommon in the environment, and possibly toxic at high concentrations. The investigators also suggested that acclimation may be more successful if it is begun with a more dilute waste stream which is less challenging to the microorganisms (Chapter 8). van Langenhove and Smet (1996) compared organic sulfide treatment in an uninoculated compost biofilter with treatment in one that had been inoculated with a culture of microorganisms enriched by growth with dimethyl sulfide. The inoculated biofilter was far more effective, with an elimination capacity of 28 g m^{-3} h^{-1} rather than 0.42 g m^{-3} h^{-1}.

If a non-compost medium is used, or when it is suspected that the compost might not contain species which degrade the contaminant of interest, general inocula are often added. Activated sludge from a wastewater treatment plant is commonly employed, because it contains an immense variety of rugged organisms which have been exposed to the typical wastes of civilization (however, concern over disease organisms that may also be present may complicate ultimate disposal of the medium). While the bulk of the species in an inoculum will die out, there are usually a few species present which will degrade the contaminant. Other general inocula are chosen with a more specific rationale. A biofilter intended to treat chlorinated hydrocarbons, for example, might be inoculated with an extract from soil at a site which has been contaminated with chlorinated hydrocarbons for many years. The presumption is that species which can degrade the compounds are likely to have become established at the site during the many years of exposure.

van Groenestijn et al. (1995) developed a biofilter for treatment of ethylene and 1,3-butadiene, using soil found at the side of a road as an inoculum. Exposed to a variety of hydrocarbons for many years, the sample contained microorganisms which degraded the contaminants well. When they started a biofilter for treating gases at high temperatures, they used an inoculum taken from compost during the high-temperature portion of the composting cycle. In both cases, nothing was known about the species present, but it was presumed that an effective inoculum would be found if a mixed culture were taken from a natural environment similar to that expected in the biofilter.

The relationship between the chosen inoculum and the characteristics of the ultimate microbial ecosystem in the steady-state biofilter are likely complex and certainly very poorly understood. General ecological observations suggest that different groups of species often combine to make ecosystems with similar structures and capabilities, and there is some indication that this occurs in biofilters. de Castro et al. (1996) compared the performance of biofilters which received three different inocula. Chipped wood mixed with mushroom compost was used to treat α-pinene after the addition of extracts from spent mushroom compost, activated sludge, or pine forest soil. Initially, the biofilter with activated sludge extract was most effective, but after 7 weeks all three were performing equally well. However, fatty acid methyl ester data analyzed by principle components analysis indicated that the biofilter inoculated with activated sludge supported different species than the other two.

In a second experiment, de Castro et al. (1997) used wood chips, compost, and perlite and inoculated with pine soil extract, pulp mill activated sludge, and municipal activated sludge to make media for treating α-pinene. Even though the cultures were grown in α-pinene-enriched liquid media for 3 weeks and were added to the biofilters at equal cell densities, differences in biofilter acclimation were seen. The pine soil extract biofilter took 4 days, the pulp mill sludge biofilter took 14 days, and the municipal activated sludge biofilter took 37 days. This was consistent with the suggestion that the degree of previous exposure the microbial consortium has to α-pinene is an indicator of how rapidly the inoculated biofilter will acclimate. In this case, too, the initial differences in biofilter performance disappeared as time passed. Fatty acid methyl ester analysis was not successful because the nature of the microbial consortium was obscured by high concentrations of fatty acids in the medium.

Webster et al. (1997) used phospholipid fatty acid analysis on the microbial ecosystems of biofilters used on wastewater plant off-gases. While the biofilters reached steady state in terms of treatment effectiveness within 100 days, the fatty acid profile was still changing in some after 300 days. Biofilters with different profiles achieved similar results. Thus, in both of these studies, ecosystems with different development histories and different species compositions functioned with equal effectiveness. The differences between them could not be seen without a sophisticated analysis.

5.3   Substrate utilization

Biofilters succeed (with the one exception of cometabolism) because there is a community of microorganisms present which use the contaminant as food, or substrate. The compound may serve as energy source or building material, or both. When it is used solely for energy, a simple organic compound is converted to carbon dioxide and water which are discharged from the biofilter. Microorganisms strive to grow and reproduce, however, and some of the carbon from the compound will often end up as part of the microorganisms. Indeed, a vigorously growing microorganism in an environment where food is abundant may convert half of its substrate carbon to biomass. In the long run, this constitutes a problem for biofilters — the biomass may accumulate and clog the reactor.

The fraction of contaminant converted to biomass is controlled by several factors. Starving microorganisms are forced to use a greater fraction of their food for energy and may have none to spare for growth. Stress may reduce the amount of growth, and some biotrickling filter operators have controlled biomass by adding salts to increase the ionic strength to stressful levels (Diks et al., 1994). Even when the contaminant-degrading microorganisms are growing rapidly, it is conceivable that biomass growth in the biofilter as a whole could be slow because the degraders are being rapidly consumed by predators.

Some compounds treated in biofilters do not contribute their elements to growth. The microorganisms in biofilters which remove hydrogen sulfide convert it almost entirely to sulfate, which remains in the water. The conversion provides energy, but very little of the sulfur is incorporated into the cells. The organism uses the energy to fix atmospheric carbon dioxide for growth.

5.3.1   Induction

The cell can only utilize a compound when the appropriate enzymes are present. The genes which code for the enzyme must be active. Some enzymes are "constitutive"; that is, they are always present in the cell and will be ready for use whenever the substrate is present. Others are "induced" and are synthesized by the organism only when the substrate appears at concentrations above an "induction threshold". Systems of this second type may limit performance in biofilters and, indeed, in all biological treatment systems. Induction will not occur if the substrate is present in concentrations below the threshold. It may also not occur if a second substrate is a better energy source because of its simpler molecular structure or higher concentration. For the biofilter as a whole, this means a low concentration pollutant may not degrade, or may not degrade in the presence of another pollutant.

5.3.2   Substrate interaction

Performance in a biofilter may be enhanced or inhibited by the interaction of the contaminants. It is often difficult to anticipate biofilter treatment success for mixed pollutants. Performance will depend on the interaction of the contaminant characteristics and the operating conditions of the system. These contaminant characteristics may include adsorptivity, solubility, bond structure, and potential biodegradability. Just as when treating pure waste streams, performance on a mixed stream is dependent on operating conditions such as loading rates, temperature, nutrient availability, moisture content, and pH. Contaminants will compete with each other for active adsorption sites in the biofilter bed, changing their availability to the microbes. The contaminants will be more or less adsorbed depending on filter medium moisture content, temperature, and pH.

Competitive inhibition may occur because the first substrate is toxic to the organisms which consume the second. Indirect ecological mechanisms may contribute. If the first substrate is present in high concentration, is readily degradable, and provides abundant energy, the organisms that use it may compete vigorously for other resources such as space and nutrients, preventing the organisms that use the second substrate from growing.

Veir et al. (1996) studied the interaction of dichloromethane and toluene. A compost biofilter was first used to treat dichloromethane and, after 53 days of acclimation, was removing essentially all of it. Addition of toluene caused an immediate decline in dichloromethane removal efficiency, but in the following 3 weeks considerable recovery was seen. The authors suggested that the species which degraded the more energy-rich toluene at first competed strongly with the dichloromethane-degrading species for another controlling resource. Over time, however, the species each dominated in spatially separate volumes within the biofilter.

Leson and Smith (1997) noted that greater time may be necessary to complete acclimation in biofilters treating complex mixtures. This suggests that the success of some species may be detrimental to others. Deshusses (1994) found that methyl ethyl ketone (MEK) and methyl isobutyl ketone (MIBK) were simultaneously removed in a compost biofilter, but when the concentration of MEK was increased the removal of MIBK decreased, and vice versa.

Generally, multiple contaminants will be adequately removed when they have similar properties. Various aromatic compounds can often be degraded by the same microbial enzymes. Several examples of biofilters treating gasoline or other mixes of benzene, toluene, ethylbenzene, and xylene (BTEX) vapors have been successfully demonstrated at both the laboratory scale and pilot scale (Chang et al., 1995, 1996; Wright et al., 1997). In some cases, the combination of similar compounds can have a positive synergistic effect on the removal of the contaminants. The presence of xylene has been shown to improve the removal rate for toluene significantly (Paca et al., 1997); however, the reverse effect did not occur. Oxygenated compounds have also been effectively removed in biofilters when they were treated in mixed waste streams.

An extreme case of negative interaction between pollutants was reported by van Langenhove et al. (1989), who observed that the removal of an undefined mixture of aldehydes was reduced from 85 to 40% by the addition of 40 ppm SO_2 to the treated air stream. Ultimately, aldehyde biodegradation was irreversibly inhibited when the biofilter was exposed to 90 ppm SO_2 for longer than 3 days.

Contaminants with very dissimilar chemical properties may still be simultaneously biodegraded in a biofilter. When overall loading rates in a biofilter are low, a diverse microbial population may thrive and utilize many contaminants. This is often the case for biofilters used at wastewater treatment facilities. In addition to the hydrogen sulfide and mercaptans found in wastewater off-gases, hundreds of organic compounds at extremely low concentrations will also be present. Effective simultaneous removal of these contaminants from such a waste stream has been demonstrated (Ergas et al., 1995; Webster et al., 1997).

5.3.3 Acclimation

It is commonly observed that microorganisms exposed to a new substrate may require a period of acclimation before they begin vigorous degradation.

There are many possible causes, and one or more of them may contribute to acclimation periods ranging from a few minutes to as much as a year. An acclimation period of a few minutes is of no consequence to biofilter operation, but if it is necessary to wait for a year before treatment begins, the system will be impractical.

The organism which is presented with a new substrate must go through a series of biochemical changes in order to begin using the compound for food. First, a chain of reactions must occur which will give the signal to "turn on" the genes which code for the needed enzymes. The genes must create the transfer RNAs, which then produce the proteins which make up the enzymes. A sufficient supply of the enzyme must accumulate. During this period, called the "lag phase", the cell often grows substantially in size without dividing. When it is fully ready, it begins to utilize the substrate and divides to produce more cells.

For common substrates, these cellular processes occur fairly rapidly, and the delay for this type of acclimation will likely be hours or days. This is commonly seen in biofiltration of petroleum fuel vapors (Leson and Smith, 1997; Wright et al., 1997). For unusual substrates, however, much longer delays have been seen. A biofilter operated to degrade methyl tertiarybutyl ether (MTBE), for example, showed no activity for a year, then suddenly became very effective (Eweis et al., 1997). Degradation may not have been possible until a cell underwent a random mutation or until a specific gene transfer occurred from one species to another. It is these processes which may take the longest times.

Even where cellular acclimation is fast, processes which might be called ecosystem acclimation may also take time. Cell division must occur for a period of time before the degrading organisms become abundant and the compound is transformed at a high rate. Competition or predation may slow the process. If the competent cells were not well distributed in the medium, it may take some time before they can colonize the whole volume. Cherry and Thompson (1997) have emphasized this form of acclimation in explanation of the observation that an early period of good treatment is sometimes followed by less effective treatment. They suggested that this could occur because the microorganisms consume substrate rapidly as they grow but are active only at maintenance levels when they reach densities at which nutrients are limiting.

Degradation may not occur immediately because no species capable of degrading the contaminant is present in the medium or inoculum. Success may be delayed until the correct organism is deposited in the biofilter on a dust particle. There is a notable difference between biofilters for air and biological systems used for water treatment. Wastewater always carries high densities of a huge variety of microorganisms. If the correct species is not present, it will surely arrive soon. Inadvertent inoculation of air biofilters, however, is much less effective. The density of microorganisms in the air is much lower, and the conditions are much less benign. The best biofilter species does well in a wet, dark environment where food is abundant. But, travel through the air exposes the cell to a completely dry environment with no food and high intensities of ultraviolet light. Many kinds of cells can produce cysts or spores in order to propagate through the air, but the ideal organism for biofiltration of a given contaminant may not be among them. The species which would be expected to arrive frequently in a wastewater treatment system might make the trip to an air biofilter only very rarely. In the MTBE work (Eweis et al., 1997), it may have been the arrival of the correct species of organism which caused degradation to start suddenly after a year of no treatment.

Acclimation to one substrate may be delayed by the presence of another. Seed and Corsi (1996) found that a biofilter could acclimate and degrade toluene alone in one day, but require 9 days to start efficient degradation when the toluene was presented as part of a mixture with benzene and xylene. This may have occurred because microorganisms were first attracted to the benzene, which is presumably more easily degraded. Only when benzene concentrations were depleted in portions of the biomass did micro-organisms begin to degrade the toluene.

Loy et al. (1997) showed that acclimation and growth of a biofilm was more rapid on polyurethane foam coated with carbon than on uncoated foam, suggesting that medium characteristics and local microenvironmental conditions may be important for acclimation rates.

Field biofilters are often shut down, either for short periods on nights or weekends or for long periods during maintenance or plant closures. Some period of acclimation may be necessary when the system is restarted, although it is generally less than that needed for a new biofilter (Deshusses et al., 1996; Swanson and Loehr, 1997). The microorganisms may be dormant but not dead, so recovery is rapid. In a biofilter treating toluene, Kinney et al. (1996) found that organisms starved for 3 days recovered in 7 hours, and organisms starved for 27 days recovered in 30 hours.

The time required for reacclimation can be reduced by caring for the microorganisms during the shut-down period. Certainly air should be provided, to avoid anaerobic conditions that will cause development of a completely different microbial ecosystem. (Some soil biofilters can be shut down each night. They are lightly loaded and less adsorptive, so that overnight degradation does not exhaust the oxygen in the medium). Water should be provided to prevent drying, which will kill many organisms. In cold climates, recovery may be more rapid if the biofilter is kept warm. Kinney et al. (1996) and Wright et al. (1997) have even recommended providing a small amount of the contaminant, so that those species which utilize it will be maintained in a healthy condition. Of course, the need for reacclimation will be reduced if variations in contaminant load can be minimized.

5.3.4   Uptake of dissolved compounds

The bulk of contaminants consumed by microorganisms in biofilters is likely taken into the cells in the dissolved form. Individual molecules are transferred from the air and exist as individual molecules in the water. Some may pass through the membrane by simple diffusion, moving from a higher concentration outside the cell to a lower concentration within. However, the cell membrane is a barrier to many compounds which may only pass when they encounter specific sites on the cell surface and are allowed through by the biochemical machinery. This control allows the cell to choose those compounds which are valuable and to exclude many which are harmful.

Some compounds are both useful to the cell and present in low concentrations in the environment, and are drawn into the cell even though concentrations within are higher. This requires that the cell have a specific catalytic apparatus in the membrane, and energy must be used to operate it. As compounds reach low concentrations in the environment around the cells, it may become energetically uneconomical for them to use the substances, and treatment will stop. For the most recalcitrant compounds, degradation may fail at concentrations that are unacceptable for release from the biofilter, and treatment will not be successful.

5.3.5   Phagocytosis

Microorganisms will often encounter substrate which is in the form of small particles. Some are capable of phagocytosis, the process in which a particle is engulfed by the cell and physically taken inside. Some protozoa use this approach to prey on bacteria and consume bits of organic matter. It is reasonable to presume that this process helps clear away organic particulate matter which is deposited in biofilters, but there have been no investigations of the phenomena. Phagocytosis is certainly important in the microbial ecology of biofilters because it is used by predators in the food chain. The protozoa used by Cox and Deshusses (1997) to reduce accumulation of biomass in a biotrickling filter were no doubt consuming bacteria in this way.

5.3.6   Exoenzymes

Some microorganisms obtain their food by releasing exoenzymes, which operate outside the cell. The fungus *Phanerocaete chrysosporium*, for example, releases a perooxidase that helps it to consume complex biological materials such as the polymers that make up wood (Braun-Lullemann et al., 1995). Lignin, hemicellulose, and cellulose are insoluble polymers and so cannot be drawn into the fungal cell as single molecules. The exoenzyme attacks the surface of the wood, breaking off short polymers and monomers which then can be absorbed by the cells. Like phagocytosis, this is a process which has been widely observed in natural environments but never specifically confirmed in biofilters. However, biofilters successfully treat some compounds which are strongly adsorbed on the support medium, and fungi are common. It is possible that adsorbed compounds are being degraded by exoenzymes.

5.3.7 Aerobic and anaerobic metabolism

The chemical reactions which cells use as energy sources, and indeed any reaction in which they process large amounts of material, must be thermo-dynamically favored. Photosynthetic organisms obtain energy from the sun, but the microorganisms which occupy the dark recesses of a biofilter must have chemical sources of energy. They support some synthetic reactions that consume energy such as those which make DNA, proteins, and other cell components, but they must carry out at least one reaction in which relatively large amounts of "food" are converted to products in a way that generates and captures energy. Any reaction that consumes energy must be used in relatively minor amounts so that the energy balance of the cell is maintained.

The energy that a compound "contains" depends on its environment. For energy-yielding reactions, microorganisms act as biological catalysts, converting compounds that are not at equilibrium with their environment to products that are. As a reaction moves toward equilibrium, it releases energy, and the cell links the reaction to processes that allow it to make use of this energy, but different parts of the environment have different background conditions. In an environment fully exposed to the atmosphere, oxygen will be abundant, and the most oxidized compounds represent the equilibrium state. Carbon dioxide and water are stable, and converting organic materials to these products will release energy. This is the case in most biofilters, in which oxygen is abundant, and the microorganisms are catalyzing the oxidation of organic contaminants. In these oxidizing environments, sulfate is the most stable form of sulfur, and microorganisms can readily survive by oxidizing hydrogen sulfide to sulfate.

Where the oxygen has been depleted, however, a reducing environment is created. Sulfate-reducing organisms can survive by converting sulfate and organic compounds to hydrogen sulfide, and carbon compounds may be converted to methane.

Thus, the chemical conversions which we can expect of the microorganisms depend on the species present and on the environmental conditions. Within a single biofilter, organic compounds may be converted to carbon dioxide and water at the surface of the biofilm, while anaerobic metabolism occurs deeper where the oxygen is depleted. If the air flow short-circuits, passing around some isolated volumes within the biofilter, they may become anaerobic. Organisms in these areas may begin to create sulfide, sulfur-containing organics, or other foul-smelling compounds, causing problems for biofilter operation.

5.3.8   Cometabolism

Most enzymes work because they have an "active site" which fits snugly on the compounds whose reactions they catalyze. An enzyme might hold two compounds together in just the right way so that a reaction occurs, and the product is released. Another compound may fit into an active site in a way which distorts its shape just enough to cause it to break apart, but molecules which are not the intended target of the enzyme may occasionally fit the same active site. The compound may be degraded even though its breakup is of no use to the organism that synthesized the enzyme. When a cell degrades a compound which provides it with no energy or other benefit, the process is called "cometabolism", and the degraded compound is the "cosubstrate". The compound that does benefit the cell is called the "growth substrate".

Cometabolism has been of particular interest with respect to chlorinated hydrocarbons. Most of these compounds do not occur in nature, and few microorganisms have developed the ability to use them as energy sources. However, the enzymes which are intended to degrade toluene or methane are sometimes effective at degrading tetrachloroethylene, trichloroethylene, or others. Many have suggested that biological treatment processes, including biofiltration of air, could be based on cometabolic processes. A reactor designed to treat tetrachloroethylene, for example, might be fed toluene as a growth substrate to keep the microorganisms healthy and to ensure an abundance of the active enzyme. Cometabolic degradation of the tetrachloroethylene would accomplish the purpose of the system.

Operation of such systems, however, has proven tricky. Speitel and MacLay (1993) have operated biofilters for treatment of trichloroethylene, using a methane feed to induce the formation of methane monooxygenase, the enzyme which also degrades the chlorinated hydrocarbon. However, when the concentration of methane was high enough to induce vigorous production of the enzyme and growth of the microorganisms, it tended to out-compete the trichloroethylene. So much of the enzyme was occupied by methane that degradation of the chlorinated hydrocarbon became slow. When the methane feed was stopped, the degradation of trichloroethylene was initially vigorous because of the absence of this competition. But, without the benefits of methane oxidation, the cells were no longer induced to produce the methane monooxygenase, and in a few hours treatment rates declined. Another period of methane feeding became necessary. It may be possible to operate a pair of biofilters in alternating sequence, so that a methane feed is rejuvenating the organisms in one while the other is treating trichloroethylene. More recently, propane and toluene have been demonstrated to be effective cosubstrates for cometabolic degradation of halogenated hydrocarbons (Ergas et al., 1993; Lackey, 1997; Sukesan et al., 1997).

5.3.9 Toxicity

High concentrations of a contaminant or a sudden increase in concentration can adversely affect microbial populations. In general, toxic shocks are more likely to occur with pollutants that have low Henry's coefficients (Deshusses, 1997). It is also possible that high concentrations may not themselves prove toxic to the microbial population, but the degradation intermediates will (i.e., because of reduced pH or high salt concentration). For hydrogen sulfide treatment in a bench biofilter, the maximum treatment capacity (above which the overloading of the system was indicated by a sudden decrease in H_2S removal efficiencies) was found to be 80 g m^{-3} h^{-1} (Allen and Yang, 1991). At higher concentrations, the hydrogen sulfide itself was toxic to the microbes. Additionally, as the hydrogen sulfide was degraded, sulfuric acid formed, reducing the medium pH. This declining pH eventually inhibited the system. Pollutant loading of more than 2 g m^{-3} of ethanol resulted in the formation of the degradation intermediates acetaldehyde, acetic acid, and ethyl acetate. The acid reduced the biofilter medium pH, inhibiting pollutant removal (Leson et al., 1993; Devinny and Hodge, 1995). Hence, care must be taken to size the biofilter correctly and to dilute the contaminants when concentrations are toxic.

Another form of microbiological toxicity may occur in a biofilter when there is a sudden introduction of a new substrate. When switching from one primary substrate to one with similar properties, a temporary lag or adjustment period may be seen. When two completely dissimilar compounds are treated in succession, toxic shock to the microbes may interrupt degradation until an adapted microbial population develops. Baltzis and Androutsopoulou (1994) abruptly switched the inlet flows of an ethanol biofilter and a butanol biofilter. Initially, poor removal of the new contaminants was observed. Over time, removal of the contaminants increased but the removal rate of butanol never equaled the rate previously recorded in the column dedicated to butanol degradation. Such results demonstrate the need for additional bench and pilot experiments before new contaminant streams are introduced to a previously operating biofilter.

5.4 The microbial community

5.4.1 Longitudinal stratification

Biofilters operate essentially as plug flow devices. Concentrations of contaminant decline as the air passes through the biofilter, so that concentrations are much higher near the influent end than at the effluent end. The characteristics of the biological community change accordingly. The influent end supports a dense biomass, possibly not substrate limited and more likely to grow until clogging occurs or until nutrients become limiting. This region will require more nutrients, generate more heat, and be more prone to acid generation and upset. The microorganisms at the effluent end may be starving and producing little or no biomass. Hugler et al. (1996), in studies of a biotrickling filter, found biofilms 5 mm thick near the inlet and only 2 mm thick two thirds of the way through.

This suggests several strategies for modified biofilter operation. If it is necessary to replace medium because of clogging, it may be possible to replace only the medium near the influent end. Some biofilters are constructed as a series of movable trays, so that trays near the influent can be treated differently. The order of the trays may be changed. A biomass tray from the influent end can be moved to the effluent end, so that the healthy biomass will provide good treatment. At the same time, the biomass will be starving, reducing clogging and head loss. Similarly, biofilters can be operated in series, with the delivery piping arranged so that it is easy to change their order in the flow stream. Farmer et al. (1995) arranged bench-scale biofilters this way and found modest reductions in head loss.

Kinney et al. (1996) operated a toluene biofilter in two phases. In the first, the biofilter was run in the usual manner until clogging had become a substantial problem in the portions of the bed near the inlet. In the second phase, the feed direction was switched every 3 days. The clogging and short-circuiting present at the end of phase one gradually disappeared during phase two. By the 71st day of alternating treatment, concentration profiles during the upflow period were essentially identical (but reversed) to those seen during the downflow period. During the first phase, 29 to 38% of the carbon entering the biofilter as toluene left the biofilter as carbon dioxide. The remainder was presumably accumulating as biomass. In the second phase, the carbon in carbon dioxide emissions rose to contain 100% and more of the incoming toluene carbon, indicating steady state or even declining biomass. While this work was done only in the laboratory, it suggests a promising strategy for biomass control. However, the authors suggest that the reactor size should be increased by a factor of two or three to accommodate the starvation periods. This may not be economically feasible.

5.4.2   Biofilms in biofilters

In microbial ecosystems everywhere, cells commonly organize themselves into films at the surfaces of solids. There are many ecological benefits to this tactic. While they may seem very active when observed in a microscope, microorganisms swim at rates which are low at the macroscopic scale. The fastest typically move only a few centimeters per minute. They are thus essentially planktonic, unable to swim against the current in even a gentle flow. A microorganism that finds itself in a desirable environment can avoid being carried away by attaching itself to some surface. Some microorganisms can "visit" the surface and leave at any time. Others attach permanently within a few minutes of their first contact.

The microorganisms become a film because many of them exude a polysaccharide gel. As the population of excreting cells accumulates at the surface, they become embedded in a continuous layer of the gel. This provides protection against predators, which often cannot penetrate the film. It also provides some protection against toxic substances, possibly because they are adsorbed on the polysaccharides. In medical practice, this is a severe problem: implanted medical devices often develop a film of infectious organisms that is protected from antibiotics.

As the biofilm thickens, it also becomes a barrier to chemical transport. The water within the polysaccharide gel is stationary, so that advection is suppressed and molecular diffusion is the only mode of transport. If the microorganisms are active, they may consume the contaminant more rapidly than it can diffuse inwards, leaving the cells deeper in the biofilm without substrate. Under these conditions the rate of treatment is controlled by the diffusion rate in the biofilm rather than by the amount of biomass present. If substrate is abundant and readily degraded, oxygen may not penetrate at a sufficient rate to supply the microorganisms, and the deeper portions of the biofilm may become anaerobic.

These chemical gradients have been the subject of extensive modeling efforts (Chapter 6) and are presumably associated with gradients in microbial species densities. Certainly aerobic species must dominate near the surface of the biofilm, while others occur where the biofilm has become anaerobic. The concentration of substrate varies with depth, and the surface species will be those capable of degrading high concentrations of contaminant, while species deeper in the biofilm can survive on lower concentrations of the contaminant, transformation by-products, or waste materials and dead cells. Mirpuri et al. (1997) proposed a model including three categories of cells: those capable of degrading toluene at high concentrations, those that can degrade toluene at lower concentrations under favorable conditions, and those that cannot degrade toluene at all. Experiments indicated that the toluene degraders are abundant near the gas-liquid interface, while the other types are abundant near the bottom of the biofilm.

Kinney et al. (1996) have noted that increasing biomass thickness may have other effects on phase transfer. A biofilm will clog any pore with an opening of maximum diameter less than twice the thickness of the biofilm. As the biofilm becomes thicker, it will fill or occlude more pores, reducing the total surface area available for gas transfer. In the extreme case, flow will be restricted to a few channels, transfer will be very slow, and the biofilter will fail.

The growth of the biofilm also reduces the total surface area available for gas transfer. Alonso et al. (1997) modeled the rate at which a growing biofilm fills the crevices near the point of contact between two support particles. A significant reduction in surface area occurs as these are filled.

In water treatment systems, biofilms are subject to "sloughing". This refers to the tendency of sections of the biofilm to break off and be carried away with the water flow. This may occur because the base of the biofilm becomes anaerobic or is starved for food, and because the shear stress from the flowing water is significant. The pieces of biofilm are collected as sludge. Efficient sloughing is vital to overall performance, because it prevents clogging of the treatment system and continually exposes new surface for more biofilm growth and the attendant contaminant consumption. Sloughing may occur in biotrickling filters but is not significant in biofilters, and so does not help to mitigate clogging there.

Biofilm models have been aggressively developed over a period of many years for reactors treating wastewater. Typically, they assume a planar biofilm of uniform thickness and model the inward diffusion of substrate and oxygen and the outward diffusion of metabolic products. In sophisticated models, the biofilm growth rate may also be simulated, in turn affecting contaminant degradation and diffusion rates. Such models have greatly advanced understanding of fixed-film treatment systems.

Many investigators have applied similar models to biofilms in biofilters used for air treatment (details are presented in Chapter 6). While this is certainly appropriate, some cautions should be observed. The biofilm in an air-phase reactor differs from that in a water-phase reactor. In the water-phase reactor, the movement of water through the porous medium exerts a considerable hydrodynamic shear. This smoothes the surface of the biofilm and carries away any microorganisms that are not attached. In an air-phase biofilter, water clings to the surface of the medium by surface tension and does not flow rapidly. Thus, it is possible for elaborate microbial structures, such as elevated fungal hyphae, to grow undisturbed. Fuzzy growth on media has been observed by many workers in real biofilters. Such growth cannot be adequately modeled as a uniform planar layer because the physical conformation of the microorganisms is so different (Figure 5.1).

Cherry and Thompson (1997) have made the observation that the number of cells in the biofilter may grow by a factor of 10^4 during the startup period. If it is presumed that the seed culture was well distributed throughout the biofilter volume, the result would be many colonies of 10,000 cells each. The colonies may produce local nutrient depletion or waste product accumulation. Some colonies may develop earlier and others later, smoothing the variation associated with acclimation and nutrient limitation.

A simple calculation can be made to demonstrate this effect. A theoretical medium of spherical particles 0.2 cm in diameter arranged in neat rows would have a specific surface area of 7.8 cm^2 cm^{-3}. If inoculum is added to provide 10^4 cells cm^{-3}, and they are 1 μm in diameter, they occupy a specific surface area of 7.7×10^{-7} cm^2 cm^{-3}. If each cell grows to produce a colony which occupies 10^4 as much area as the original cell, the total area covered is 0.77 cm^2 cm^{-3}, or about 10% of the total. Thus, for most of the acclimation period, the cells exist as individual colonies, and completion of a biofilm only comes when cell counts are very high.

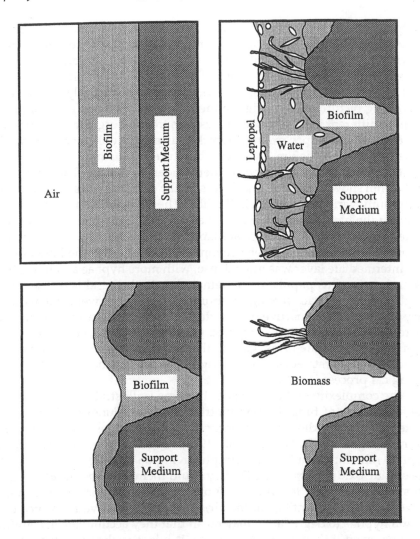

Figure 5.1 Various conceptual models of biomass in biofilters.

A more realistic picture of biofilm structure suggests the possibly important phenomena which are ignored in traditional models. Because the water phase in the air biofilter is relatively stationary, it can support a population of swimming microorganisms. These will be free to move near the surface of the water, and will see higher concentrations of substrate because of the lesser diffusion limitation, and may contribute significantly to the activity of the biofilter.

Hugler et al. (1996) measured bacterial concentration in the biofilm and water phases of a biotrickling reactor used for treating sulfides. They found the bacterial density in the water to be about one third that in the biofilm. They suggested that this was not an important factor for treatment in biotrickling filters; however, it remains notable that a substantial fraction of the cells was suspended in the water. The stationary water in a biofilter may hold more. The same investigators developed a particularly complete and interesting picture of the biofilm in their biotrickling filter. The biofilm may be quite different in biofilters, or in systems treating organic substances, but this is a well-developed example which illustrates biofilm complexity. They characterized the biofilm as a "vast mixture" containing fungi, bacteria, yeasts, ciliated protozoa, amoebae, nematodes, and algae. Their sections of the biofilm showed three distinct layers, as the characteristics of the microbial community varied with depth in the film. The external layer, at the interface between the biofilm and the water phase, was a thin whitish zone of dense bacteria and the tips of the fungal filaments, or hyphae. There were pockets observed which may have been channels which allowed the flow of water. The intermediate layer was more dense, with more hyphae and fewer bacteria and a vigorous population of nematodes. In the basal layer, next to the solid surface, hyphae were again abundant, but many were collapsed as if they were dead or inactive. Other dormant fungi were seen, with a few bacteria. The investigators suggested that this deeper layer may have been deprived of substrate or oxygen, so the organisms were not healthy. They noted with understatement that "... rigorous mathematical modeling of this biological process is not trivial ... ."

The complexity of the ecology of biofilms is illustrated by discussion of the abundance of fungi in their biotrickling filter. Fungi might not be expected in a biotrickling filter treating inorganic sulfides, which are rapidly degraded by bacteria. Hugler et al. (1996) suggested four possible explanations: (1) the fungi could be autotrophs living on organic matter created by the sulfide oxidizing bacteria, (2) they could themselves be sulfide oxidizers (this is observed in a few fungi), (3) organic substrate for the fungi may be present in small quantities in the air, and (4) the relative resistance of the fungal hyphae to degradation may mean that they accumulate in significant amounts even if they grow very slowly. While it is clear that the fungi are important organisms in this biotrickling filter, there is still a fundamental lack of knowledge of their function in its microbial ecosystem.

The erect filamentous organisms may grow into the air space. Their hydrophobic surface remains dry, and they absorb substrate directly from the air stream. The high surface area of the filament can greatly facilitate phase transfer. Braun-Lullemann et al. (1995) suggested the importance of direct adsorbtion of contaminants on the filamentous white rot fungus used in a biofilter.

The convoluted surface of the realistic biofilm may also provide extra surface for mass transfer. Indeed, some investigators (Potera, 1996) have seen pores and channels within biofilms and suggest that the microbial community organizes itself in this manner in order to promote transfer of substrate and oxygen from the water phase by advection rather than diffusion.

Modeling the effects of such intricate biofilm microstructure currently exceeds our ability. Direct observation is difficult because of the delicacy of the biofilm. Drying the material destroys its shape, and it cannot be easily sectioned for microscopic study. Models which assume simple planar biofilms still provide the best insight possible; however, their limitations should be kept in mind, and efforts to make them more realistic should continue.

5.4.3 Higher organisms in biofilters

The dominant pollutant degraders in biofilters are bacteria and fungi. These simple organisms are capable of utilizing the substrate rapidly. Their small size produces a high surface-to-volume ratio ideal for rapid pollutant uptake. Bacteria are generally smaller and more active than the fungi, and they will dominate in a biofilter with a high water content treating easily degraded substrate at near neutral pH. Fungi will become more important if the biofilter is drier or more acidic, and they can degrade some complex substances which are beyond the metabolic abilities of bacteria.

As biofilters are open to colonization by new species of bacteria and fungi, they are also open to colonization by higher organisms. de Castro et al. (1996) made microscopic observations of three biofilters which had received inocula from various sources. In the system inoculated with activated sludge, ciliated protozoa were observed on day 26 of operation, rotifers on day 54, and nematodes on day 70. All of these are common in sewage sludge. In a column inoculated with soil extract, nematodes were observed sooner. Thus, both the biofilters developed longer food chains, but the nature and timing of the development depended on the inoculum.

Insect larvae may be seen. The compost used in many biofilters will bring with it a host of worms and insects, and many of these may thrive in the system if the pollutant is not too toxic. In some cases, the insect larvae have metamorphosed to flying insects, which have been a minor nuisance. In one closed biofilter, the upper headspace was in turn colonized by thousands of spiders which were preying on the flying insects.

While little careful study has been done, all of this suggests that biofilters can support complex ecosystems with long food chains. Flying insects may occasionally be an irritation, but in general the establishment of these food chains is desirable. As the predators consume the bacteria and fungi, they use much of what they eat for energy and use only a small part of the biomass for their own growth. Typically, more than 90% of the biomass which is eaten will be converted to carbon dioxide and water. At least 90% of the predator biomass will be degraded when they are eaten by their predators, so the organic material which enters the food chain is rapidly eliminated from the biofilter. It is likely that this has a substantial influence on reducing clogging

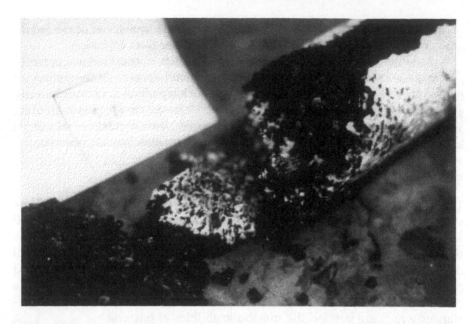

Figure 5.2 Excessive microbial growth can cause clogging and channeling. This is the granular activated carbon medium removed from a bench-scale biofilter. It was removed partially intact because the heavy growth had cemented the particles together. The white areas are solidly clogged with microbial growth; the longitudinal black channel was the last remaining avenue for air flow. This ethanol biofilter was heavily overloaded. (Photograph courtesy of J.S. Devinny.)

in biofilters where a low, relatively non-toxic load allows a complete ecosystem to develop.

Cox and Deshusses (1997) utilized this phenomenon in control of biomass in experimental biotrickling filters. Two biotrickling filters used for degradation of toluene were identical except that an inoculum of various species of protozoa was added to one. The biotrickling filter which received the protozoa had less biomass, a lower pressure drop, and better treatment efficiencies after 80 days of operation.

5.5   Biomass clogging

Biomass accumulates in a biofilter when growth from the introduced organic carbon exceeds endogenous respiration. When loads on the system are large and mineral nutrients are abundant, the biomass may clog the filter bed packing material. Media clogging can produce large pressure drops and form air channels (Figure 5.2). Back pressures on the blower equipment increase wear on the system and raise electrical demand. Air channeling will limit the amount of contaminant being treated and will negatively affect the performance of a biofilter. Excessive growth may also lead to other, more chronic problems. Examples include enhanced deterioration of the filter bed packing material by the large consortium of microbes and filter bed compaction as biomass weight builds. As clogging increases, anaerobic zones of activity may develop and odorous end products may be generated. Limiting excessive biomass growth is essential for biofilter success.

Numerous studies have focused on predicting excessive microbial growth and devising means to control clogging. Biomass growth is commonly concentrated near the inlet of the biofilter. As the biofilm thickens at the inlet because of the higher concentrations of organic substrate, clogging conditions begin to prevail. This clogging causes treatment effectiveness to decline near the inlet of the biofilter, and further growth occurs deeper in the reactor. Research on one biofilter found 10^{10} viable heterotrophic bacteria per gram of medium at the inlet of the filter bed, and 10^8 viable bacteria per gram of medium at the outlet (Medina et al., 1995a). The growth of microorganisms along the length of the filter bed will generally be proportional to the removal rates. Zero-order kinetics may be seen at the inlet of the system, while first-order pollutant removal may occur deeper in the medium.

Clogging control strategies have ranged from rudimentary methods, such as mechanical tilling or water injection, to more advanced methods of nutrient limitation or biomass starving through flow reversal. In the earlier days of biofilter reactor operation, the predominant method of biofilter growth control was to remove the material, rotate or till it, and place it back into the reactor. This approach works well for small biofilters, but becomes costly and labor intensive for larger units. The addition of water or backwashing at high flow rates may be effective in shearing off biomass from the packing material, but should only be used on inorganic media (Sorial et al., 1994, 1997). Application of flowing water to organic filter beds is not advisable because of the probability of compaction, leaching of nutrients, and enhancement of air channeling.

Nutrient limitation may control biomass effectively on inorganic filter beds. Because the stoichiometry of biomass is relatively constant, reducing the total nutrient concentration limits the amount of biomass that can form. Various studies have looked at controlling nitrogen or phosphorus concentrations for this purpose, with good results (Morgenroth et al., 1995; Smith et al., 1996; Weber and Hartmans, 1996; Govind et al., 1997). However, maintaining nutrient concentrations at a level that will limit biomass growth with minimum damage to treatment efficiency is a difficult balancing act.

Switching the direction of air flow in a biofilter has proven to be one effective means to control excess growth without long lag times in performance (Kinney et al., 1996). The initial flow direction (either top or bottom loaded) provides a carbon-rich area near the inlet and a carbon-deprived area at the outlet. After some time, the flow is reversed, reversing the carbon-rich and -deprived areas as well. However, starving a large portion of the biofilter by overdesigning the reactor might not always be economical.

For all clogging control methods, some limitations exist. Biofilter media may be clogged in some portions of the bed, but not in others. It may be necessary to perform "smoke" or tracer test experiments on the system to determine if clogging conditions are contributing to channeling problems. If clogging is evident from such tests, removal of the medium may be warranted.

5.6   Microbial observation of biofilters

To fully understand the pollutant removal mechanisms in a biofilter, thorough analysis of the microbial population performing the degradation is warranted. Many methods exist for both quantitative and qualitative observations of microorganisms and their activities. Most are not specific to biofilter studies but have been developed over many years in clinical microbiology, microbial ecology, and waste treatment and have been adapted to characterizing biofilter microbial samples.

5.6.1   Microscopic observation

Microscopic observation is the simplest means of establishing a general picture of the microbial community on a biofilter medium sample. Numerous microscope configurations with various magnification abilities exist. The light microscope is convenient for most routine use, while the various electron microscopes are used for more extensive research purposes. Both types of microscopes may be used for quantitative and qualitative information. However, for the more advanced microscopes, increased training and complicated sample preparations are required. Techniques for more advanced microscopy are detailed elsewhere (Brock and Madigan, 1991).

5.6.1.1   Light microscopes

Direct microscopic counts of microorganisms may be performed to enumerate different species in a biofilter sample (Brock and Madigan, 1991; Atlas and Bartha, 1993). Sample preparation requires placing a known mass of biofilter medium in a known volume of buffering solution, mild vortexing and sonication of the sample to ensure that the microorganisms are transferred to the liquid phase, and counting of the microbes under a light microscope using a counting chamber. The counting chamber is marked with grids of known volume within which microbes can be visually counted. The counts are converted to microbes per milliliter of sample. Because biofilter samples consist of large quantities of organic matter, it may be necessary to stain the microorganisms in order to differentiate them from detritus. Some fluorescent stains will mark living organisms but not dead cells or detritus. This can make direct counts far more effective. If the sample is extremely turbid, a series of diluted samples may be prepared. Knowing the original sample volume and dilution ratio, the investigator can calculate the density of microbes on the biofilter medium.

The direct microscopic count technique is valuable but has many limitations. Dead cells may not be distinguishable from live cells. Cells may move from grid to grid, confusing the count, and small cells are difficult to detect under the microscope. Detritus may obscure cells or be mistaken for cells. Most microorganisms appear as featureless circles, ovals, or other simple shapes, so species cannot be identified. Still, the technique may provide important information.

The light microscope can also be effectively used for a qualitative assessment of the microbial population in a biofilter. It may be used to identify larger prokaryotic and many eukaryotic species from their appearance. Gram stain may be employed to classify microbes according to their structure. The staining effect differs with the makeup of the cell walls of the microbes and allows typical species to be classified by their ability to retain the stain. The culture is stained with a crystal violet/iodine dye and washed with an alcohol-based decolorizer. Organisms in which the dye is removed are called gram negative. Gram-positive organisms are those for which the dye does not wash out, so they appear blue under the microscope. The tendency to stain is associated with fundamental differences in the characteristics of the cell wall. Separating the microbes into the simple classes of gram-positive and gram-negative will improve the overall understanding of the type of microbes that inhabit the biofilter. In general, gram-negative organisms are fast growing, utilize many carbon substrates, and adapt quickly to a wide variety of environments. Gram-positive organisms are generally slower growing than gram-negative organisms, are capable of degrading more recalcitrant compounds, and are more resilient to changing environmental conditions. Overall, the light microscope is a valuable tool which can supply general quantitative and qualitative information about the microorganisms inhabiting a biofilter.

5.6.1.2 Electron microscopes

Used primarily as a research tool, the electron microscope can provide crucial quantitative and qualitative information about the microbial community on the biofilter media. The biomass on individual particles can be mapped. From such precision, important factors such as filter media biofilm coverage, thickness, and activity can be determined for use in modeling. The electron microscope can prevent mistakes in estimating such parameters. Using transmission electron microscopy, Cox (1995) described how the active biomass treating styrene was mainly present at the surface of the medium (observed as cells with intact cytoplasm), whereas the cells in the deeper parts of the biofilm were dead (observed as cell wall remnants). An overestimation of biofilm thickness might have been made had the electron microscope not been utilized. Additionally, the biofilm itself was irregular in structure, indicating that biofilm thickness varied over the surface of the medium. These irregularities should be accounted for when determining biofilm thickness for modeling purposes.

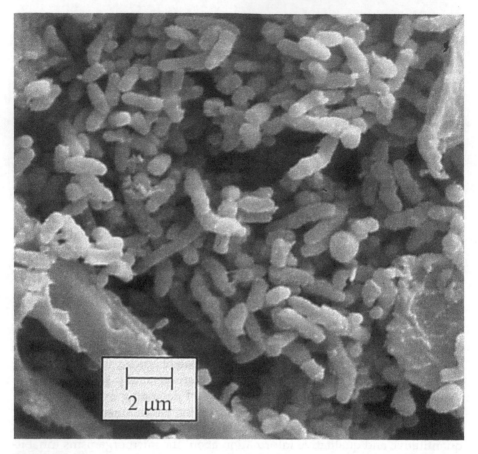

Figure 5.3 Scanning electron micrograph of a biofilm formed after 90 days of biofilter operation (original magnification 5000×). Toluene is the substrate, and peat is the support. Biomass growth is quite vigorous. (Photograph courtesy of Villanueva, C., Acuna, M.E., Auria, R., and Revah, S., 1997).

The electron microscope can also be used as a qualitative tool to identify microbial structures such as mycelia, spores, and individual cells. These may indicate the presence of certain classes of microorganisms in the biofilter. These groups may not be discernible using the light microscope or other identification techniques. The electron microscope can also determine characteristics of the filter material that are important to the system's success. If the filter bed is showing increases in pressure drop or poor performance because of channeling, electron microscopy may show whether clogging is arising from accumulation of biomass or from aggregation of small medium particles. (See Figures 5.3 to 5.5.)

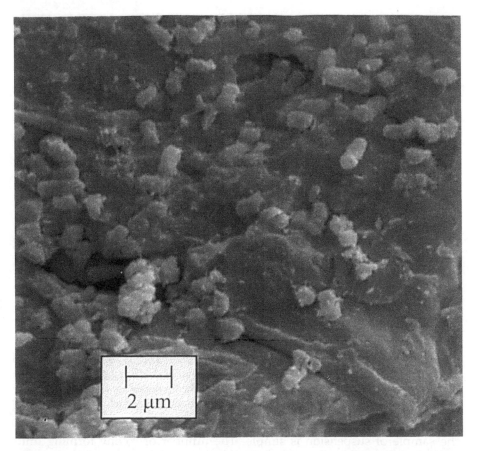

Figure 5.4 Scanning electron micrograph of a biofilm formed after 90 days of biofilter operation (original magnification 5000×). Toluene is the substrate, and peat is the support. The initial nutrient concentration was 10% of that for the biofilm shown in Figure 5.3. Biomass growth is evident, but is much less vigorous. (Photograph courtesy of Villanueva, C., Acuna, M.E., Auria, R., and Revah, S., 1997).

5.6.2 Viable heterotrophic plate counts

The viable heterotrophic plate count is a convenient and inexpensive microbial enumeration technique (Alexander, 1977; Brock and Madigan, 1991). When a sufficiently dilute suspension is added to a nutrient agar plate, individual cells grow to become distinct colonies on the surface of the agar. Different species produce colonies of different colors and shapes. Fungal colonies may have a fuzzy appearance. It is presumed that each colony on the plate arises from a single cell in the original suspension. The colonies can be seen and counted with the naked eye, so determining the number of cells in

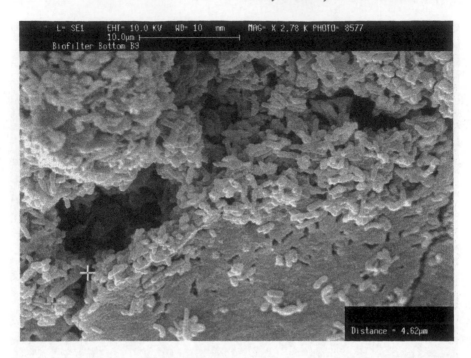

Figure 5.5 Scanning electron micrograph of microorganisms on activated carbon medium from a biofilter used for anaerobic treatment of chlorinated hydrocarbons in simulated landfill gas (Devinny et al., 1997). Note the tendency for microorganisms to cluster around the pore openings. (Photograph courtesy of Schwarz, B. and Tsotsis, T.)

the sample of suspension is simple. The number of colony forming units (CFUs) can be counted and expressed as the number of microbes per gram of wet or oven-dry medium.

Plate counts can provide a second kind of information through the use of selective growth media. Only organisms that grow are counted, and in many cases, the culture medium can be chosen so that only the organisms of interest will appear. For example, if an investigator wishes to count the number of organisms likely to be useful in a biofilter treating ethanol vapors, agar can be prepared in which the only substrate is ethanol. Those species that grow and are counted must be ethanol degraders. If only a count of bacteria is desired, a fungicide can be included in the agar.

In many preparations, including those from biofilter samples, it is likely that there will be large numbers of cells present. The plate inoculated directly with such a sample will have thousands or millions of overlapping colonies, too numerous to count and impossible to distinguish separately. It is usually necessary to prepare a dilution series by placing a known quantity of biofilter medium into a known amount of a sterile buffering solution. Subsamples of this suspension are then transferred to fresh solution to dilute the microorganisms. Continued dilutions are performed until 30 to 150 microorganisms will appear on a nutrient agar plate after incubation. Counts lower than 30 will not be statistically significant, while more than 150 colonies will be difficult to count because of possible overlapping.

Though the viable heterotrophic plate count technique is convenient and inexpensive, numerous limitations exist and many assumptions are necessary. Foremost, the viable plate technique only counts those microbes that can grow on the specific nutrient agar used. It is quite possible that a large percentage (up to 99%) of the microorganisms which are active in the biofilter sample are not culturable, producing erroneous population estimates (Amann et al., 1995). Species that were dormant in the biofilter may grow well in the nutrient agar, suggesting that a species is active and important when it is not. Additional assumptions are made about the homogeneous nature of the liquid sample. It is assumed the vortexed dispersion is homogeneous with microbes and that all microorganisms are adequately transferred from the medium sample to the liquid. It is very possible that microbes may be aggregated. If several cells stick together, they will only form one colony on the plate, which will be assumed to represent only one cell. Therefore, adequate vortexing and sonication of the sample should be done in the hope of separating all the microorganisms from the solid medium and providing uniformity in the suspended phase.

Even with its limitations, the technique still provides a valuable base of knowledge about the general number of microbes in the biofilter. Typical bacterial numbers from biofilters range from 10^8 to 10^{10} per gram of sample, while fungal populations may be 10^3 to 10^6 per gram of sample (Medina et al., 1995a). If many samples are taken over the life of the biofilter project, counts will provide valuable information about the general health and status of the microbial population.

5.6.3 Phospholipid fatty acid analysis

Lipids are an important constituent of all microorganisms. In conjunction with proteins, they form a highly selective barrier, the cell membrane, which enables the microbe to concentrate specific nutrients and excrete waste products. The type and composition of lipids, as well as their elemental form of fatty acids, are unique to different species of microorganisms. This uniqueness may allow for identification of microbial species in a medium sample. Fatty acid methyl ester (FAME) analysis is a simple method used to analyze the fatty acids of the lipids. The analysis begins with saponification and methylation of the cells, followed by extraction of the FAMEs, which are dissolved in a solvent and analyzed by gas chromatography. The chromatogram represents a snapshot of the mixed microbial community. The chromatogram can be used for monitoring changes in the microbial community over time.

Table 5.1 Microorganisms Associated with Certain Classes of Fatty Acids

Class of fatty acid	Associated microorganism
Terminally branched saturates	Gram-positive and sulfate-reducing bacteria
Monoenoics	Gram-negative and some microeukaryotes
Polyenoics	Eukaryotic microorganisms
Branched monoenoics	Sulfate-reducing bacteria
Mid-chain branched saturates	Gram-positive and sulfate-reducing bacteria

Source: Vestal, J.R. and White, D.C., *Bioscience*, 39(8), 535, 1989. With permission.

A major drawback of the FAME process is the background interference caused by fatty acids not related to the microbial communities. Because the method extracts all the fatty acids in the sample, it is very difficult to match fatty acids with a specific group or type of microbes. de Castro et al. (1997) found that FAME analysis was effective in distinguishing between microbial communities from different sources, but detailed analysis was hampered by interference. Results obtained using FAME can be more specific if microbes from a biofilter medium sample are isolated in cultures of individual species before FAME analysis is performed. This minimizes the amount of interference caused by other species and fatty acids, but also greatly increases the time and difficulty of the procedure. Species that cannot be cultured in the lab will be missed.

Extraction and subsequent analysis of the phospholipid fatty acids (PLFA) that make up the lipid bilayer of the cell membrane can provide qualitative and quantitative information on the microbial populations. PLFA analysis has been predominantly used for pure cultures, but recent advances in the technique for mixed cultures have proven useful for biofilter samples (Webster et al., 1997).

In PLFA analysis, the crude lipids of a sample are extracted using a monophasic mixture of chloroform, methanol, and water (Bligh and Dyer, 1959). Silica gel chromatography separates the crude lipids into polar lipids, neutral lipids, and diglycerides. Methanolysis and dimethyl disulfide (DMDS) derivatization prepare the polar lipids for analysis by gas chromatograph/ mass spectrometry.

The PLFA analysis provides living biomass estimates and indications of community structure, metabolic status, and ecological stress for microbes in a biofilter sample. Live bacteria retain a relatively fixed amount of PLFA in their cell membranes. In dead cells, the phospholipids are quickly hydrolyzed to neutral diglycerides. The total polar fraction of PLFA can be used to estimate living biomass. Community structure may also be inferred by identifying certain fatty acids that serve as signature lipid biomarkers (SLBs). These SLBs aid in identifying certain broad classes of microorganisms (Vestal et al., 1989; see Table 5.1). Metabolic status and stress factor information is also provided by specific SLBs (Guckert et al., 1986a,b; Vestal et al., 1989; White et al., 1995).

The PLFA technique for observing microbial populations provides a good overall picture of microbial population characteristics. However, the analysis is labor intensive, requires highly specialized personnel and training, and utilizes sophisticated, expensive instrumentation.

5.6.4 *Deoxyribonucleic acid (DNA) extraction*

One of the most advanced techniques for biofilter microbial observation is deoxyribonucleic acid (DNA) extraction and analysis. Using the polymerase chain reaction (PCR), DNA sequences can be "amplified" or copied in great numbers. A probe DNA molecule causes multiple amplification products to form, creating a "fingerprint" that can be used to identify various species of microorganisms. This procedure allows for a detailed investigation of microbial communities in a biofilter. However, because the probes are specific for individual microbial strains, some general prior knowledge about the species in the biofilter sample is needed.

While DNA extraction is a well-established technique in clinical and microbiology, its use in biofilter applications has been limited (de Castro, 1997). Obtaining meaningful results from heterogeneous microbial populations in biofilters is difficult, and the laboratory equipment and training required to run such experiments are expensive.

5.7 *Conclusions*

A biofilter is only successful when the microbial ecosystem it contains is healthy and vigorous. Proper operation of a biofilter, to a considerable extent, consists of maintaining this biological activity. It is important to remember that the microorganisms are part of an ecosystem, subject to environmental conditions and interactions among the species.

Bohn (1996) has emphasized this point in comparing biofilters using artificial media and complex engineered control systems with simpler reactors using compost or soil. He notes, "Minimizing size by maximizing control can create instability. Microbial populations fluctuate with time for reasons which are difficult to determine. Microbes release growth inhibitors and bactericides which slow and stop microbial activity. These problems are accentuated by smaller size. A system with more internal self-regulation is likely to function better longer."

This philosophy deserves close consideration, especially when it is noted that elaborate engineering control is expensive and sometimes unreliable, and natural controls are free. Control of the biofilter environment may allow process optimization, a small footprint, and economical operation, but only if the operator knows what the optimal environmental conditions are. Ecosystems consist of large numbers of organisms working diligently to make efficient use of resources. When that resource is the contaminant being fed to a biofilter, allowing an ecosystem similar to those found in nature to flourish in its own way may be a very effective approach.

Whether biofilters reproduce ecosystems found in nature or create highly artificial ones, it is important to remember that the organisms present will continue to follow the laws of ecology. It is a milieu in which simple rules combine to govern very complex systems, and a fundamental understanding of the system will contribute to good biofilter management.

chapter six

Modeling biofiltration

6.1 Introduction

Beginning with the early development of biofilters, efforts were directed toward modeling. The objectives were to organize experimental data and to understand simple relationships between parameters such as media surface area, biological activity, biofilm thickness, etc. and pollutant removal. In addition, a real interest exists in biofilter modeling for design purposes, i.e., being able to predict the performance of a biofilter under given conditions. Finally, biofilter models can also be used for process optimization.

Biofilter modeling started in the early 1980s and was based on earlier work on submerged biofilm models (Jennings et al., 1976; Ottengraf, 1977). The models assumed basic mass balance principles, simple reaction kinetics, and a plug flow air stream. One of the questions to be answered with these models was the value of the overall order of reaction (first-, zero-, or half-order) occurring in biofilms (Harrenmoes, 1977). Extensions of these submerged biofilm theories by Ottengraf et al. (1983) became the basis for future biofilter modeling. Ottengraf's model (1986) is still most commonly referenced, and many other models were developed from it. For this reason, this model is explained in greater detail in the next section. Recently, with increases in computing power, the trend has been to develop more complex models or to use "brute force" to solve a large number of equations. Some of the most significant biofilter models are reviewed in the next sections. Even more recently, a fundamentally different but potentially promising type of model has been developed (Choi et al., 1996; Govind et al., 1997; Johnson and Deshusses, 1997). These models use quantitative structure activity relationships (QSARs) and seek to predict the performance of biofilters from data describing the removal of a few known pollutants.

6.2   The challenge of modeling biofiltration

The difficulty in modeling a biofilter lies in the complexity of the fundamental processes. Biofiltration involves many physical, chemical, and microbiological phenomena. In order to simulate biofilter effectiveness with varying operating conditions, a model must include these various phenomena. Further, a number of unknowns or difficulties exist in the definition of equations for a biofilter model. These include:

- Biofilters are relatively ill-defined systems; there is no agreement in the scientific community on their operation principles.
- Biofilters are tubular reactors, and often very different conditions prevail at the top and the bottom of the reactor. This means that pollutant biodegradation has to be integrated from one end of the reactor to the other. Therefore, if other changes are considered (e.g., within biofilm depth), two directions for integration must be simultaneously performed. In the lab, this problem has sometimes been overcome by using a differential biofilter, where part of the outlet air is recycled to the inlet, thereby establishing homogeneous conditions throughout the reactor.
- A good understanding exists of the kinetic relationships describing the substrate biodegradation rate by suspended pure cultures, but, in biofilters, biodegradation is mediated by mixed cultures immobilized in a biofilm, and little is known about the intrinsic microkinetics in biofilters. Expressions for degradation kinetics may differ significantly for those two systems (Møller et al., 1996; Mirpuri et al., 1997; Pederson et al., 1997).
- Microscopic observation of biofilms is a difficult task. Good physical descriptions of biofilter media surfaces and of biofilms do not exist at this time. In the absence of such information, a flat geometry is most often assumed for the biofilm.
- In common biofilter studies, only gas-phase concentrations are measured. It is very difficult, if not impossible, to measure experimentally the physico-chemical conditions that are actually experienced by the process culture. Thus, it is necessary to make assumptions that are often impossible to verify. This makes it difficult to determine values for the model parameters, and to validate model development.
- Sorption of pollutants by heterogeneous damp material is poorly understood. Here, again, extrapolations have to be made.
- Using Fick's second law requires solving a second-order differential equation. The introduction of a biological kinetic, e.g., Monod or Michaelis-Menten, makes it impossible to solve the model equations analytically.
- If a biofilm type of model is assumed and numerical techniques are used to solve model equations, difficulties exist in applying boundary

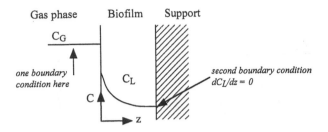

Figure 6.1 Problem caused by the nature of the boundary conditions when modeling a biofilm in a biofilter.

conditions. One boundary condition sets the contaminant concentration in the gas phase, and a second boundary condition dictates the concentration gradient at the substratum interface (it should be zero; see Figure 6.1). Both conditions refer to different geometrical extremities of the problem, and common numerical integration may require trial and error to simultaneously satisfy the two conditions.

- Experiments with biofilters may be difficult to compare to one another, as they can be sensitive to the experimental procedures.
- In a biofilter, EBRTs are on the order of minutes, sorption usually takes hours to days, and biomass growth to steady state may take months. A model which efficiently represents all these time scales can be very computationally intensive.
- The output data that are used to determine whether the model is valid are quite simple, consisting of single removal or perhaps a concentration profile within the biofilter. Steady-state profiles are typically straight lines or exponential curves. Because these simple forms can be reproduced by many different models, it may be difficult to determine which model is valid.

In the following sections, different models are briefly reviewed. Their general characteristics are summarized in Table 6.1.

6.3 Biofilm models

6.3.1 Ottengraf's model

This model was first published in 1983 (Ottengraf et al., 1983). The best description available is found in Ottengraf (1986). Ottengraf's model is based on developments made for the biodegradation of non-adsorbable substrates in submerged biofilms by Jennings et al. (1976). Ottengraf modified this model to accommodate gas/liquid biofilms. He also included the study of three basic situations expected to be common in biofilters for waste air treatment, i.e., first-order kinetics, zero-order kinetics with reaction rate

Table 6.1 Summary of the Principal Features of Selected Biofilter Models

Model	Section	Characteristics	Solution	Validation on	Design applicability
Ottengraf, 1986	6.3.1	Simple steady-state model, assumes first or zero-order kinetics	Analytical	Compost/peat biofilter	Possible with some restrictions due to the assumption of the kinetics
Devinny and Hodge, 1995	6.3.2	Dynamic model, assumes uniform solids/water phase	Numerical, finite differences	Carbon and compost biofilters	Possible for some aspects
Shareefdeen et al., 1993	6.3.3	Steady-state model with oxygen limitation and substrate inhibition	Numerical, trial and error for the biolayer concentration profile	Peat moss perlite biofilter	Possible/difficult
Shareefdeen and Baltzis, 1994	6.3.4	Transient model with oxygen limitation; patches of biomass	Quasi-steady state, numerical solution	Peat moss perlite biofilter	Possible/difficult
Deshusses et al., 1995a,b	6.3.5	Dynamic model, includes sorption and cross-inhibition between two pollutants	Numerical, finite difference	Compost biofilter	Possible/difficult
Morgenroth et al., 1995b	N/A[a]	Dynamic model, includes growth and decay of the process culture	Numerical	JP5 removal in GAC biofilter	Uncertain

Reference	Section	Description	Type	Application	Extrapolation
Cherry and Thompson, 1997	N/A[a]	Dynamic model, includes nutrient limitation, cell growth and pollutant consumption for cell maintenance purposes	Numerical	Mixed hexanes, compost biofilter	Uncertain
Zarook et al., 1997	N/A[a]	Dynamic model, includes sorption and cross-inhibition between two pollutants	Quasi-steady state and general numerical solution	Peat moss perlite biofilter; removal of benzene and toluene	Possible/difficult
Choi et al., 1996	6.4.1	Model for comparing treatment of several pollutants, model for the first-order rate constant	QSAR with various parameters	Compost and activated carbon biofilters	Possible in the future; extrapolation still uncertain
Johnson and Deshusses, 1997	6.4.2	Model for one biofilter and treatment of various VOCs, model for EC_{max} and $Load_{95\%}$	QSAR with group contributions and various parameters	Compost biofilter	Possible in the future; extrapolation still uncertain
Govind et al., 1997	6.4.3	One differential biofilter, fitting of Monod parameters	QSAR, with group contribution only	Compost biofilter	Possible in the future; extrapolation still uncertain

[a] Model not described in the text. Interested readers are referred to the published paper.

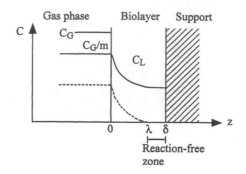

Figure 6.2 Geometry considered by Ottengraf for his biophysical model of the biofilter.

limitation, and zero-order kinetics with diffusion rate limitation. The model is based on the following assumptions (Figure 6.2):

1. The gas-phase interfacial resistance is negligible; hence, interfacial concentration can be assumed to be in equilibrium with the gas phase concentration.
2. The gas phase is in plug flow, i.e., no radial concentration gradients exist.
3. Pollutant transport in the biofilm is by diffusion and can be described by an effective diffusion coefficient, D_{eff}.
4. The biofilm thickness, δ, is small relative to the support particle diameter, so that a flat geometry can be assumed for the biofilm.
5. The microkinetics for substrate elimination in the biofilm can be described by a Michaelis-Menten type expression.

$$k = k_{max} \frac{C_L}{K_M + C_L} \tag{6.1}$$

Further, the kinetics can be simplified to either first-order kinetics ($C_L \ll K_M$) or zero-order kinetics ($C_L \gg K_M$).

Many of these assumptions are quite reasonable, others are quite restrictive. The first one can be justified by empirical correlation for gas/liquid mass transfer (Thoenes, 1958). Plug flow behavior can be assessed experimentally by measuring the dispersion of a pulse of an inert tracer within a biofilter. The analysis of the residence time distribution demonstrates that in most cases, for biofilters operating under standard conditions (EBRT of 0.5 to 2 min.), plug flow is occurring. A typical residence time determination is

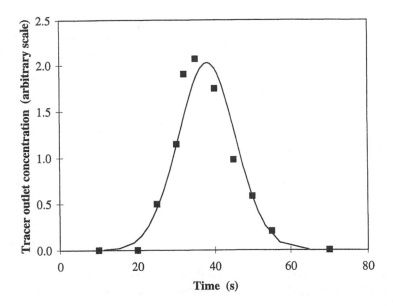

Figure 6.3 Residence time distribution in a biofilter; the tracer pulse of methane was injected in the inlet stream at time zero. Bed height: 92 cm; EBRT: 75 s. The outlet gaseous concentration (symbols) is plotted vs. elapsed time, and the line represents fitting with a dispersion model. A dimensionless Bodenstein number of 52 was calculated. Reactors with Bo numbers larger than 20 are considered to be plug flow reactors. Others have found Bodenstein numbers ranging from 40 to 200 in biofilters (Sabo, 1991). (From Deshusses, M.A., Biodegradation of Mixtures of Ketone Vapours in Biofilters for the Treatment of Waste Air, Ph.D. thesis, Swiss Federal Institute of Technology, Zurich, 1994.)

shown in Figure 6.3. Analysis of empty bed residence time may also allow detection of short-circuiting (channeling) or dead zones. However, the study of residence time distribution is not easy on full-scale biofilters, in particular in open biofilters. In such cases, injection of smoke in the inlet air is a convenient alternative for detection of short-circuiting.

The third assumption made by Ottengraf is reasonable, but, as far as the value of the effective diffusion is concerned, great uncertainties (up to 2 orders of magnitude) exist (Siegrist and Gujer, 1985; Characklis and Marshall, 1990; Christensen and Characklis, 1990). Biofilms differ widely, and experimental measurement of effective diffusion in biofilms is very difficult. The diffusion coefficient might even vary with the depth in the biofilm, due to accumulation of exopolymers produced by the process culture. The fourth model assumption (flat biofilm geometry) is most probably wrong (see Chapters 2 and 5), but because there is no "universal biofilm geometry", using a planar geometry is a lesser of two evils. The last assumption refers to the

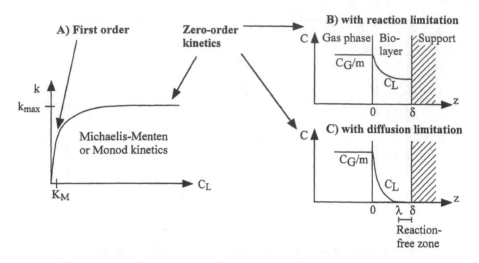

Figure 6.4 The choice of one of the three situations for the kinetic expression considered by Ottengraf in his model is made on the basis of concentration in the biofilm.

biodegradation kinetic of the pollutant, and postulates a Michaelis-Menten type of microkinetics, with pollutant carbon as the only limiting substrate. As discussed in Section 6.3.3, this might not be true when the pollutant is not the only limiting substrate. Other models presented further in this chapter or published recently have more complex kinetics, such as multiple substrate competition or oxygen limitation.

The Ottengraf model seeks to differentiate among three possible operating situations, i.e., first-order kinetics, zero-order kinetics with reaction rate limitation, and zero-order kinetics with diffusion rate limitation. The principal reason for this is that it allows an analytical solution of the differential equations. It is an important limitation of the model, because two or three of these situations may very well be encountered in the same biofilter but at different locations. The simplification of the kinetics for the three situations and the procedure to solve model equations are summarized in Figures 6.4 and 6.5.

The results for the gas-phase pollutant concentration with respect to filter height for first-order kinetics, zero-order with reaction limitation, and zero-order with diffusion limitation kinetics, respectively, are given by:

$$\frac{C_G}{C_{Gi}} = \exp\left(-\frac{hK_1}{mv_a}\right) \qquad (6.2)$$

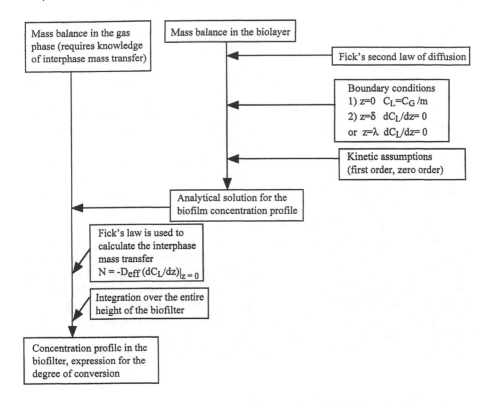

Figure 6.5 Schematic of the setup and solution of Ottengraf's model equations.

$$\frac{C_G}{C_{Gi}} = 1 - \left(\frac{hK_0}{C_{Gi}v_a} \right) \tag{6.3}$$

$$\frac{C_G}{C_{Gi}} = \left[1 - \frac{h}{v_a} \sqrt{\frac{K_0 D_{eff} a}{2mC_{Gi}\delta}} \right]^2 \tag{6.4}$$

where C_G = gaseous concentration; C_{Gi} = inlet gaseous concentration; h = height in the biofilter; K_0 = zero-order reaction rate constant; K_1 = first-order reaction rate constant; m = the gas liquid partition coefficient; v_a = the superficial velocity; D_{eff} = the effective diffusion coefficient; δ = the biolayer thickness.

From here, it can be concluded that the concentration profile along the height of the biofilter is exponential, linear, or quadratic for first-order, zero-order with reaction limitation, and zero-order with diffusion limitation, respectively. It should be noted that, in the first case, complete removal is theoretically impossible, and, in the case of high removal with zero-order reaction, the concentration of pollutant will actually drop to such extent that first-order kinetics should theoretically apply.

For this model, the interfacial area is calculated from the average radius of the biofilter media, and the microkinetic parameters are obtained from shake flask experiments, where the pollutant is supplied to cultures isolated from the biofilter. It is not certain that this accurately reflects conditions in the biofilter, because the suspended microbial ecosystem in shake flasks and the system immobilized in biofilters might not have the same characteristics.

Comparison of experimental results obtained in a laboratory compost/peat biofilter for the elimination of common VOCs and model simulation are shown in Figure 6.6. The effect of diffusion limitation can easily be seen at inlet concentrations of 0.2 to 0.3 g m^{-3} for butyl acetate, and at about 1 g m^{-3} for toluene. The effect of diffusion limitation is also illustrated in Figure 6.7, where the experimental concentration profile in the biofilter is compared to the theoretical profile for both the reaction limitation (straight line) and the diffusion limitation cases. Validation of the Ottengraf model was achieved by other authors for various conditions (van Lith et al., 1990; Dharmavaram, 1991).

6.3.2   Devinny and Hodge model

Devinny and Hodge developed a model that made simple assumptions about conditions in the water and on the medium in order to emphasize the effects of changing inlet concentrations (Devinny et al., 1991; Hodge and Devinny, 1995, 1997). The model assumed that transport in the water and adsorption on the medium are rapid in comparison to advection and biodegradation. Under these conditions, contaminant concentration in a phase consisting of the solids and water (the solids-water phase) can be described as a single number which is the mass of contaminant in the solids-water phase divided by the volume of the phase. This ignores the distribution of contaminant within the water and medium, where concentrations are higher at the surface of the water than at the bottom of the biofilm and are zero within the medium solids. Further, biodegradation is assumed to follow a first-order kinetic and thus is equal to a constant multiplied by the solids-water phase concentration.

In this model the volume of the biofilter consists of two phases — the air and the solids-water — with concentrations defined in each. At equilibrium, the ratio of the concentrations can be described by a quasi-Henry's constant or concentration partition coefficient k_h:

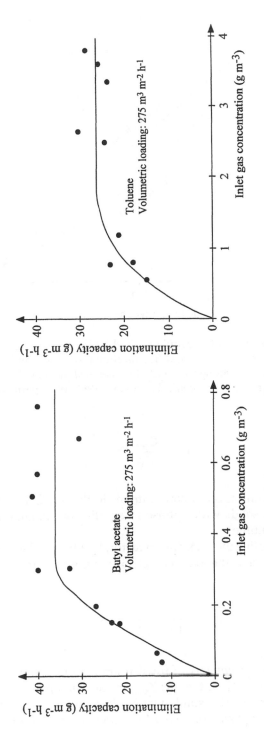

Figure 6.6 The elimination capacity of a biofilter as a function of the inlet gas concentration, model (line), and experimental data (symbols). (From Ottengraf, S.P.P., in *Biotechnology*, Vol. 8, Rehm, J.J. and Reed, G., Eds., VCH Verlagsgesellschaft, Weinheim, 1986. With permission.)

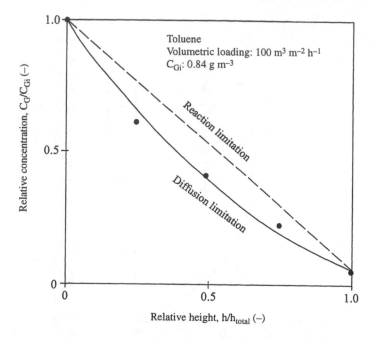

Figure 6.7 Toluene concentration profile in the biofilter: experimental data (symbols), model simulations in the case of diffusion limitation (line), and reaction limitation (dashed line). (From Ottengraf, S.P.P., in *Biotechnology*, Vol. 8, Rehm, J.J. and Reed, G., Eds., VCH Verlagsgesellschaft, Weinheim, 1986. With permission.)

$$k_h = \frac{C_{SW}}{C_G} \tag{6.5}$$

where C_G = the concentration of contaminant in the air; C_{SW} = the average concentration in the solids-water phase at equilibrium (contaminant mass divided by phase volume).

It is possible to define the retardation factor, R, which is the ratio between the velocity of the air and the average velocity of the contaminant (unitless; see Chapter 2):

$$v_{cont} = \frac{v_i}{1 + K_{mass}} = \frac{v_i}{R} \tag{6.6}$$

where v_{cont} = the average velocity of the contaminant (length per unit time); v_i = the interstitial velocity of the air; K_{mass} = the mass partition coefficient (mass of contaminant in the solids-water phase divided by the mass of contaminant in the air phase for a small volume of the biofilter).

The equation for the concentration of contaminant in the air includes terms for longitudinal dispersion, advection, and transfer to the solids-water phase:

$$\frac{\partial C_G}{\partial t} = D_L \frac{\partial^2 C_G}{\partial x^2} - v_i \frac{\partial C_G}{\partial x} - \left(\frac{1-\theta}{\theta}\right)[k_t(k_h C_G - C_{SW})] \qquad (6.7)$$

where t = the elapsed time; D_L = longitudinal dispersion coefficient; x = longitudinal distance within the biofilter; θ = porosity (volume of air per unit volume of biofilter); k_t = transfer rate coefficient (per unit time).

The rate of mass transfer is presumed proportional to the degree to which the phase is below saturation ($k_h C_G - C_{SW}$) and to the transfer rate constant, which depends on the available surface area, turbulence, temperature, and other factors. To convert mass transfer rates to rates of change in concentration, it is necessary to multiply by the ratio of the volumes of the solids-water phase and the air phase.

A second equation describes concentrations in the solids-water phase as increased by transfer from the air and decreased by biodegradation:

$$\frac{\partial C_{SW}}{\partial t} = [k_t(k_h C_G - C_{SW})] - b C_{SW} \qquad (6.8)$$

where b is the first-order biodegradation rate constant.

The transfer term is the negative of the phase transfer term for the air, and biodegradation is assumed linear with the concentration of contaminant in the solids-water phase. This implicitly assumes that all of the contaminant in the solids-water phase is available for biodegradation.

The model is easily solved by finite differences, and can be applied to arbitrary initial conditions for concentrations along the length of the biofilter. Arbitrary input concentrations can be used as a boundary condition at x = 0. Thus, while it presents an oversimplified view of phenomena within the water phase and biofilm, it is convenient for investigating the effects of changes in input concentrations. In particular, it will readily model a "pulse test", in which a spike of contaminant is added to the biofilter, forming a concentration peak which moves down the biofilter at a rate determined by the retardation factor and grows smaller at a rate determined by the biodegradation rate constant.

If it is assumed that the system is at equilibrium with respect to phase transfer, and the definition for the retardation factor is used, the equation for the concentration in the air becomes:

$$\frac{\partial C_G}{\partial t} = \frac{D_L}{R} \frac{\partial^2 C_G}{\partial x^2} - \frac{v_i}{R} \frac{\partial C_G}{\partial x} - \frac{b K_{mass}}{R} C_G \qquad (6.9)$$

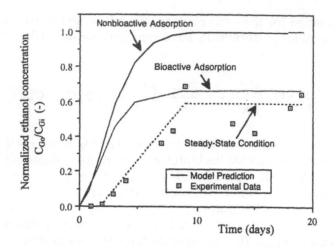

Figure 6.8 Comparison of breakthrough profiles for experimental data (symbols) and model predictions (full lines) in a compost biofilter. The dotted line represents the least square approximation of the data. (From Hodge, D.S. and Devinny, J.S., *J. Environ. Eng.*, 121(1), 21, 1995. With permission.)

If it is further assumed that input concentrations are steady, longitudinal dispersion is negligible, and the system has reached steady state, there is an analytical solution to Equation 6.9:

$$C_G = C_{Gi} \exp\left(\frac{-bK_{mass}x}{v_i} \right)$$ (6.10)

where C_{Gi} = reactor inlet air concentration; C_G = concentration along the height, x, of the biofilter.

Thus, concentrations within the biofilter decline exponentially, with a rate constant proportional to the biodegradation rate constant and the mass partition coefficient and inversely proportional to the air flow velocity. The biofilter works better if biodegradation is rapid and if the contaminant partitions well into the solids-water phase. Many studies have indicated that biodegradability and solubility are the two most important factors determining the relative success of a given biofilter for various contaminants.

Two examples of model simulations and experimental results are shown in Figures 6.8 and 6.9 for biofilter treating ethanol. Figure 6.8 shows a breakthrough profile, i.e., the biofilter outlet concentration vs. time after startup in a compost biofilter. The upper line shows the adsorption pattern only (b set to zero in Equation 6.9). Figure 6.9 shows ethanol and carbon dioxide concentration profiles in a granular activated carbon (GAC) biofilter.

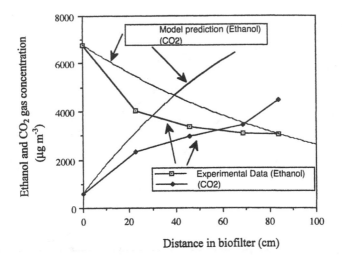

Figure 6.9 Experimental data and model fitting for ethanol and carbon dioxide concentration profiles for a GAC biofilter. (From Hodge, D.S. and Devinny, J.S., *J. Environ. Eng.*, 121(1), 21, 1995. With permission.)

6.3.3 *Shareefdeen et al. model*

Another biofilter model was published by Shareefdeen et al. in 1993. It is based on assumptions similar to those of Ottengraf, with differences in the expression for microkinetics. Here, a double limitation by both the carbon substrate and oxygen is considered, and it is assumed that at least one of the rate-limiting substrates (methanol or oxygen) is depleted before it reaches the biolayer/solid interface. This is similar to defining an effective biolayer thickness, λ, as discussed by Williamson and McCarty (1976). Substrate self-inhibition at high concentrations is also included. The resulting Haldane type kinetic is given below. Also, numerical integration is performed, avoiding the choice between first- or zero-order kinetics. These are all valuable improvements because both changes in the kinetics within the reactor and oxygen or substrate limitation are likely to occur in reality.

$$k = k_{max} \frac{C_{L,j}}{K_{M,j} + C_{L,j} + (C_{L,j}^2 / K_{i,j})} \frac{C_{L,O}}{K_{M,O} + C_{L,O}} \tag{6.11}$$

The model uses a number of parameters, such as the growth rate, interfacial area, biofilm density, etc. Some of the values for these parameters were determined experimentally, while others were taken from previously published work. In the 1993 paper, the model was validated for different experiments where either flow or methanol inlet concentrations were varied

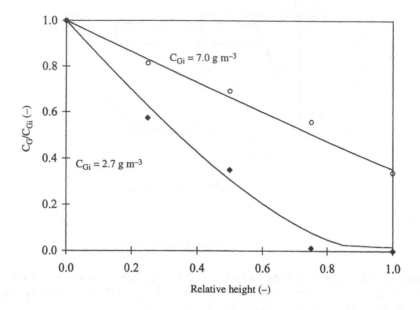

Figure 6.10 Comparison of experimental (symbols) and modeled (lines) concentration profiles for methanol removal in a biofilter. EBRT: 3.0 min. (Courtesy of Shareefdeen, Z., 1998.)

(Figures 6.10 and 6.11). The question of constant interfacial area and biomass density was raised. Some of the concentration profiles could be better fitted toward the bottom of the column if a reduced interfacial area was chosen, which would indicate that bed compaction and channeling, i.e., a change in interfacial area, did indeed occur.

Interestingly, the thickness of the effective biolayer changes with the concentration and, thus, with position in the biofilter. As expected, the effective biolayer thickness increased as the concentration of methanol decreased. According to Shareefdeen et al., in all cases where methanol gas concentrations were larger than 2.7 g m^{-3}, oxygen was the first substrate to be depleted. But, the model showed that both methanol and oxygen had to be taken into account in the microkinetic expression.

6.3.4   *Shareefdeen and Baltzis model with patches of biomass*

In an extension of their previous model exposed in Section 6.3.3, Shareefdeen and Baltzis (1994) refined their theory and introduced sorption of the pollutant to the packing and partial coverage of the support medium by biofilm in order to be able to simulate transient changes occurring in biofilters. As in their previous model, an effective biolayer was assumed in which either the primary pollutant (toluene) or oxygen was depleted. The difference between

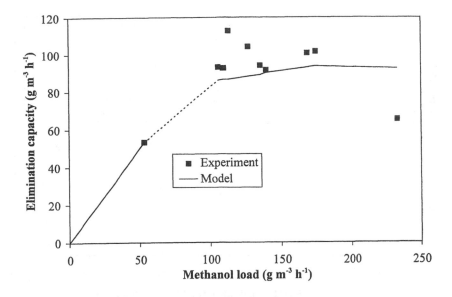

Figure 6.11 Comparison of experimental data (symbols) and model simulation (line) for a number of experiments. Compiled with data from Shareefdeen et al. (1993).

the two models is in the nature of the coverage of the support particles by the biolayer. Here, patches of biofilms are assumed (see Figure 6.12), leaving bare surfaces of the solid in direct contact with the air stream. This allows direct adsorption of the pollutant on the solid particle. This adsorption was described in the model by a Freundlich isotherm. For solving the equations, because the problem is very stiff, a quasi-steady state was assumed and numerical approximation was performed. Model parameters were determined either by fitting of biofilter experiments or in separate experiments with biofilter medium or with suspended cultures in shake flasks. A key parameter is α, the fraction of the total interfacial area covered by the biofilm.

α = Biofilm area / Total interfacial area

Figure 6.12 Schematic of the physical situation modeled by Baltzis and Shareefdeen. (Adapted from Shareefdeen, Z. and Baltzis, B.C., *Chem. Eng. Sci.*, 49, 4347, 1994. With permission.)

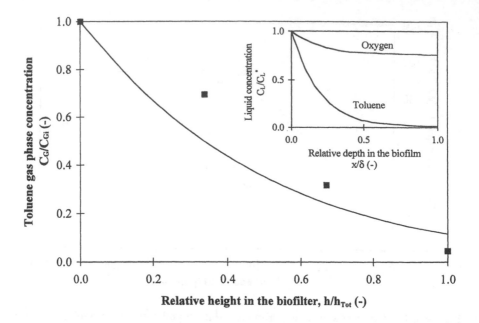

Figure 6.13 Steady-state experimental (symbols) and modeled values (curve) for the elimination of toluene in a peat/perlite biofilter. Volumetric loading was 7.8 m³ m⁻³ h⁻¹. The inset shows the model-predicted dimensionless biofilm concentration profile of oxygen and toluene at mid-biofilter height. It shows that in this case, toluene was depleted before oxygen in the biofilm and thus determined the effective biofilm thickness. (Adapted from Shareefdeen, Z. and Baltzis, B.C., *Chem. Eng. Sci.*, 49, 4347, 1994. With permission.)

A value of 30% was found to best describe most biofiltration experiments, but no experimental confirmation was obtained for this value. Even so, the model has some significance in that it is the first to go beyond uniform planar geometry. Figures 6.13 and 6.14 show some of the experimental results and how they compared to simulated values.

6.3.5 *Deshusses et al. model*

Deshusses et al. (1994, 1995a,b) developed the first model which described the transient behavior of biofilters and the details of diffusion processes within the biofilm. The motivation for a dynamic model was that in biofilters, transient conditions are the rule rather than the exception, and that a number of very interesting phenomena occur during the transients.

As a starting point, transient events, such as the response of a biofilter after a change in the concentration or air flow were modeled using a traditional biofilm model. The model results showed that a steady state should be obtained within a few seconds. Such a result is definitely lower than what is

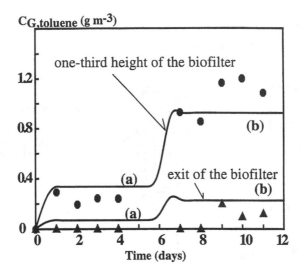

Figure 6.14 Experimental data (symbols) and model prediction (curves) for a step-change in operating conditions. Conditions were changed from a toluene inlet concentration of 0.68 g m^{-3} and an EBRT of 8.6 min. (condition a) to an inlet concentration of 1.65 g m^{-3} and an EBRT of 7.7 min. (Adapted from Shareefdeen, Z. and Baltzis, B.C., *Chem. Eng. Sci.*, 49, 4347, 1994. With permission.)

generally observed experimentally (several hours). The reason is that sorption affects biofilter dynamics. Sorption isotherms of methyl ethyl ketone (MEK) and methyl isobutyl ketone (MIBK) on biofilter media could be well approximated by gas/damp volume equilibrium (Deshusses, 1994); therefore, it was assumed that a sorption volume made of water dispersed in the support matrix existed and that no biological reaction would occur in this sorption volume. This assumption was used to split the biofilter conceptually into sections. The fate of the pollutants was described by mass balances for each section (Figure 6.15). Conceptually, the polluted air flows downwards so that convection is the vector of pollutant transport in the gaseous phase. At the gas-biofilm interface, equilibrium is assumed, i.e., gaseous and interfacial liquid concentrations are related by Henry's Law. In the biofilm, the pollutants simultaneously diffuse and are consumed by the microorganisms. Storage of the pollutants in the sorption volume is also possible, but only after their diffusion through the entire thickness of the biofilm (Figures 6.15 and 6.16).

A number of assumptions similar to those of Ottengraf were made. The most significant are

1. Each subdivision, as defined in Figures 6.15 and 6.16, is ideally mixed, so that its concentration is homogenous.

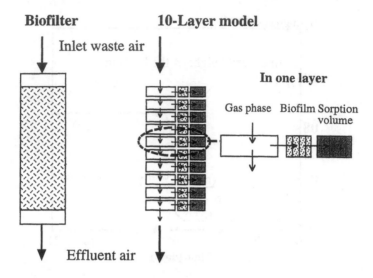

Figure 6.15 Stage-wise model of biofilters showing the structure used for finite-difference calculations. Biofilters are divided into ideally mixed subdivisions. (From Deshusses, M.A. et al., *Environ. Sci. Technol.*, 29(4), 1048, 1995. With permission.)

2. The sorption volume is assumed to be equal to the water content of the support material minus the biofilm volume, and no biological reaction takes place within the sorption volume.
3. Oxygen limitation does not occur (this was confirmed experimentally), but mass balances for oxygen could be incorporated if deemed necessary.
4. In the biofilm, no net growth of biomass is assumed so that kinetic parameters remain constant over the time considered. Michaelis-Menten type of kinetics with competition between substrates are assumed to follow Equation 6.12.

$$k_j = \frac{k_{max,j} C_{L,j}}{K_{m,j}(1 + C_{L,i} / K_i) + C_{L,j}} \qquad (6.12)$$

where k_j and $k_{max,j}$ are the biodegradation rate and maximum biodegradation rate, respectively, of compound j per unit of biofilm volume; j refers to the pollutant evaluated, and i to the competing substrate; K_i = competitive inhibition constant; $C_{L,i}$ or $C_{L,j}$ = contaminant concentration in the biofilm.

For every subdivision, a dynamic (i.e., dC/dt) mass balance equation was written and solved by finite differences. Typically, 3000 to 5000 iterations were necessary to simulate 5 to 10 hours of biofilter operation. Interestingly,

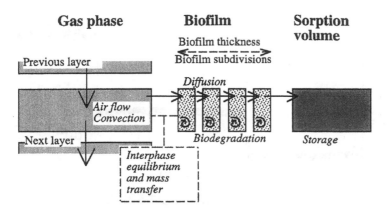

Figure 6.16 Schematic description of the model for one layer. The polluted air flows through the gaseous section, gas-biofilm equilibrium is assumed, and simultaneous diffusion and biodegradation of the pollutants occurs in the biofilm subdivisions. No biodegradation of the pollutants occurs in the sorption volume. (From Deshusses, M.A. et al., *Environ. Sci. Technol.*, 29(4), 1048, 1995. With permission.)

writing the model in the dynamic form avoids the boundary condition dilemma discussed in Figure 6.1. This is because the system will converge to a solution that is stable and satisfies the boundary conditions. This stable value can be used as a steady-state solution. The model uses a number of parameters, such as interfacial area, biofilm thickness, degradation rate, etc. These parameters were calibrated in a set of biofiltration experiments specifically designed for this purpose. The model was then validated on independent experiments (Deshusses et al., 1995b), or used for model parametric sensitivity (Deshusses et al., 1995a).

Experimental results were compared to the model prediction (Figure 6.17). In this experiment, the air flow rate was kept constant and the inlet concentration of MEK was increased stepwise. The model was also used to simulate steady performance (Figure 6.18). Elimination characteristics were shown for both the simultaneous treatment of equal concentrations of MEK and MIBK and for treatment of MEK alone. As discussed in Chapter 4, significant reduction of performance can occur when mixtures of pollutants are treated simultaneously in a biofilter. For the first time, this was included in a model, and, as shown in Figure 6.18, a fair description of the reduction of the performance resulting from the presence of multiple pollutants was obtained.

Finally, an interesting simulation of Deshusses et al. (1995b) is the study of the hypothetical variation of conditions during the treatment of MEK and MIBK vapors. The simulated case considered an active biofilter restarted after an interruption of several days, with the inlet concentration of one

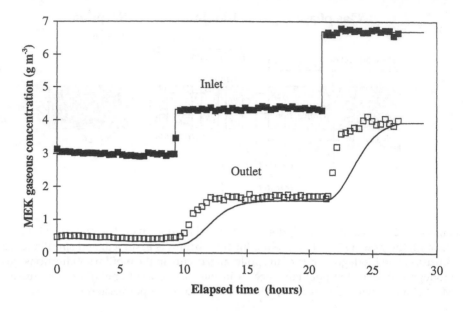

Figure 6.17 Dynamic response of a biofilter to step changes in MEK inlet concentration during MEK removal as a single pollutant at a volumetric loading of 44 m³ m⁻³ h⁻¹. The solid line represent model predictions; the symbols are experimental data. (From Deshusses, M.A. et al., *Environ. Sci. Technol.*, 29(4), 1059, 1995. With permission.)

pollutant, MIBK, kept constant and the concentration of the other pollutant, MEK, increased in a stepwise manner and subsequently disconnected (Figure 6.19). Predicted steady state is reached after 3 hours. At this time, complete removal of MEK and 80% removal of MIBK are predicted. After the restart, the dynamic increase of pollutant penetration and the marked effect on concentration profiles of competition between MEK and MIBK can be seen after 0.5, 1, and 6 hours, respectively (Figures 6.19 and 6.20). After 10 hours, the MEK inlet concentration was increased stepwise. Whereas the model predicts a breakthrough of MEK, 95% of the MEK was removed. Significant repression of MIBK degradation was predicted, resulting in an increase in the MIBK outlet concentration from 0.1 to 0.4 g m⁻³. Corresponding modifications in the concentration profiles occurred after 11 and 18 hours, respectively (Figure 6.20). In the simulated case, MEK was interrupted after 20 hours. However, persistence of MEK in the outlet stream for as long as 2 hours after interruption was predicted as a direct consequence of a combination of desorption and biodegradation. Thereafter, MIBK was removed as a single pollutant, undergoing complete elimination, contrary to the first 10-hour phase where its breakthrough was induced by the inhibition of MEK. Significant differences could be observed in the MIBK concentration

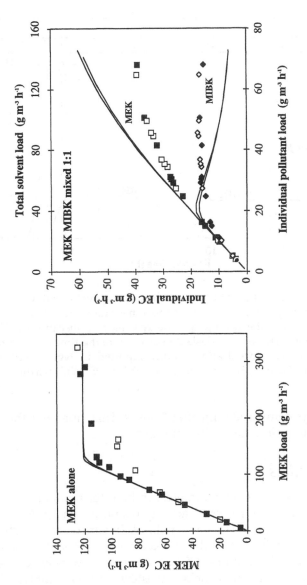

Figure 6.18 Comparison of experimental measurements (symbols) and model-predicted biodegradation/loading characteristics (lines), for MEK removal as a single pollutant, or when mixed with an equal concentration of MIBK in the inlet air. Open symbols: experiments at 44 m^3 m^{-3} h^{-1}; full symbols: experiments at 88 m^3 m^{-3} h^{-1}. (From Deshusses, M.A. et al., *Environ. Sci. Technol.*, 29(4), 1059, 1995. With permission.)

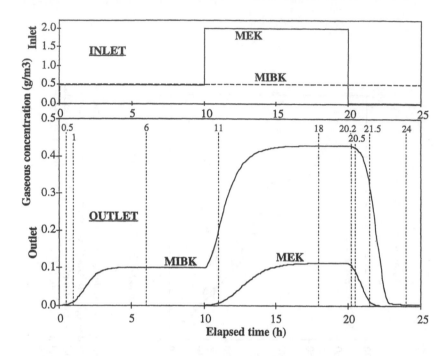

Figure 6.19 Dynamic simulation of the simultaneous removal of MEK and MIBK vapors from a waste air stream in a biofilter. In the upper graph, step changes in inlet concentrations are shown. In the lower graph, the dynamic response predicted by the biofilter model is shown. The vertical dashed lines indicate the times chosen for the evaluation of the biofilter concentration profiles reported in Figure 6.20. (From Deshusses, M.A. et al., *Environ. Sci. Technol.*, 29(4), 1048, 1995. With permission.)

profile after 24 hours compared with after 6 hours. This illustrates the major impact of competing pollutants on one another.

6.4   QSAR models

Recently, there has been a renewed interest in biofilter modeling with the development of quantitative structure activity relationships (QSARs) (Choi et al., 1996; Devinny et al., 1997; Govind et al., 1997; Johnson and Deshusses, 1997). QSAR models seek to describe the activity of particular chemicals based on their chemical structure. QSARs have been widely applied in toxicology and more recently in the field of wastewater treatment (Govind et al., 1991; Boethling et al., 1994; Okey and Stensel, 1996). The application to biological waste air treatment is new and shows promise for biofilter design. QSAR models are quite different and in some ways much more limited than the conceptual models. Because the only data used are those that describe the compound, QSAR models obviously cannot describe all aspects of biofiltration.

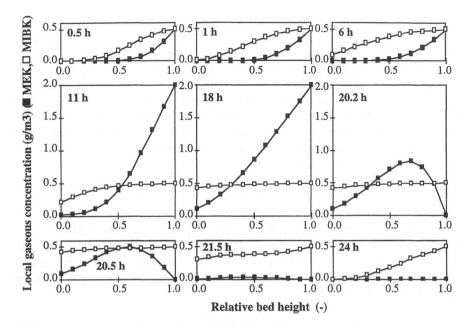

Figure 6.20 Evolution of concentration profiles in the biofilter with respect to time for simultaneous removal of MEK and MIBK corresponding to Figure 6.19. The symbols (MEK ■, MIBK □) are the modeled local gaseous concentrations in the different sublayers of the model. Because of the down-flow mode of operation, the highest local pollutant concentrations occurred at a relative bed height of 1 (inlet), and the lowest occurred at a relative height of 0 (outlet). Times correspond to the vertical dashed lines shown in Figure 6.19. (From Deshusses, M.A. et al., *Environ. Sci. Technol.*, 29(4), 1048, 1995. With permission.)

What can be expected of QSAR models is that they will describe the relative treatability of various compounds for a set of conditions. Thus, where there is experience with a set of compounds in a biofilter, it should be possible to predict the performance of that biofilter or a similar biofilter on other compounds.

6.4.1 Choi et al. model

Choi and Devinny (Choi et al., 1996; Devinny et al., 1997) developed a QSAR model of biofiltration using data collected for two compost and three granular activated carbon biofilters used on wastewater treatment plant off gases. Removal efficiencies were measured for seven chlorinated hydrocarbons, four aromatics, and five aldehydes or ketones. The project investigated molecular connectivity indices and valence molecular connectivity indices ($^0\chi$, $^0\chi^v$, $^1\chi$, $^1\chi^v$, $^2\chi$, $^2\chi^v$, $^3\chi$, $^3\chi^v$, $^3\chi$, and $^3\chi^v$) and logarithms of molecular weight, octanol-water partition coefficient, Henry's Law constant, and loading rate per unit volume. Molecular connectivity indices for each pollutant are easy to calculate knowing the chemical structure. These indices have been used extensively in toxicology, drug research, and other medically related applications. Biofilter treatment performance, indicated by the ratio of outgoing concentration to incoming concentration, was used as the dependent variable for multiple regression. Hence, the correlations obtained are restricted to the specific conditions used for the test set of data. Regressions were done for the results for individual columns, averaged results for GAC columns, averaged results for compost columns, and overall average results.

No single parameter was effective at predicting treatment success. Where two parameters were used, however, several relationships had squared Pearson correlation coefficients of about 0.8. In many cases, the results for one column, used as a training set, predicted the results of the other columns used as test sets. K_{ow} and $^3\chi^v$ were the best predictors for GAC columns, while K_{ow} and $^3\chi$ were best for compost and for overall averages. The overall relationship had a squared correlation coefficient of 0.895, showing that the fit was quite good:

$$C_{Go} / C_{Gi} = 0.300 - 0.435 \ ^3\chi + 0.241 \log K_{ow} \qquad (6.13)$$

This is presumably valid for biofilters operating at an EBRT of about one minute on gases of low concentrations (1 ppb to 10 ppm).

The importance of K_{ow} in biofilter performance is expected. Hydrophobic compounds have less tendency to dissolve in the water phase and so are carried more rapidly through the biofilter by the moving air. It is notable that for five of the chlorinated hydrocarbons and three of the carbonyl compounds, $^3\chi$ is zero. Among these, K_{ow} alone predicted relative treatment success. It may be that the primary limiting factor for treatment of these is their solubility. Because biofilter success is largely controlled by compound solubility and biodegradability, it is tempting to suppose that $^3\chi$ represents biodegradability. However, for this data set, it was not related to the known degradability of the compounds, and a physical interpretation of its significance is not obvious.

6.4.2 Johnson and Deshusses model

Johnson and Deshusses (1997) considered individual removal data for 16 VOCs in a laboratory-scale, compost-wood chips biofilter and sought to describe either the maximum elimination capacity (used to design biofilters for a maximum throughput) or the load at which 95% pollutant removal occurred (high removal design). These two values specific to each pollutant correspond to the two extremes in the elimination capacity vs. load curve discussed in Chapter 1. The Johnson and Deshusses model used the dimensionless Henry's Law coefficient, the logarithm of the octanol-water partition coefficient, a molecular connectivity index, and group contributions. The essential relationship is described in Equations 6.14 and 6.15 (EC_{max} and $Load_{95\%removal}$ are in g m^{-3} h^{-1}).

$$Log(EC_{max}) = \alpha H + \beta \log K_{ow} + \gamma^1 \chi^v + \sum_{groups} n_{group} \delta_{group} \qquad (6.14)$$

$$Log(Load_{95\%removal}) = \alpha H + \beta \log K_{ow} + \gamma^1 \chi^v + \sum_{groups} n_{group} \delta_{group} \qquad (6.15)$$

where α, β, γ, and δ_{group} are model parameters (different in Equations 6.14 and 6.15) to be calibrated.

The rationale for the choice of H, K_{ow}, $^1\chi^v$, and group contribution descriptors is that the elimination of a pollutant in a biofilter is likely to be influenced by gas-liquid partition and by biodegradation rate or toxicity of the treated pollutant. In the model, these properties are described by the Henry's Law coefficient, and both the octanol-water partition coefficient and the chemical structure, respectively. The influence of the chemical structure is split between a molecular connectivity index $^1\chi^v$, which is easy to calculate from chemical bonds and valence electrons (Kier and Hall, 1986), and group contributions δ_{group} representing the different groups in the treated pollutant (e.g., CH_3, OH, aromatic CH, etc.). The removal data for 15 different VOCs was used to calibrate the model parameters α, β, γ, and the different δ for either Equation 6.14 or Equation 6.15. A summary of the results is shown in Figure 6.21 where a good agreement between data and model can be observed. It is likely that in the future, such a model will be used for reactor design purposes.

6.4.3 Govind et al. model

Govind et al. (1997) investigated the removal of various volatile organics in a micro-batch biofilter (see Figure 6.22) packed with inert celite pellets and operated with a minimum of trickling nutrient solution. Because the air was recycled in a batch manner at a high rate, a so-called differential biofilter was established. This allowed investigators to neglect gas phase mass transfer resistance and made the determination of kinetic parameters much easier because uniform concentrations could be assumed over the height of the biofilter. Govind developed a set of equations allowing the decrease of the pollutant with time shown in Figure 6.22 to be described with the two parameters of a Monod kinetic (μ and K_s) and constants specific to the system.

The removal of toluene, trichloroethylene, methylene chloride, chlorobenzene, and ethylbenzene was then used to determine a group contribution for μ and K_s as formulated in Equations 6.16 and 6.17.

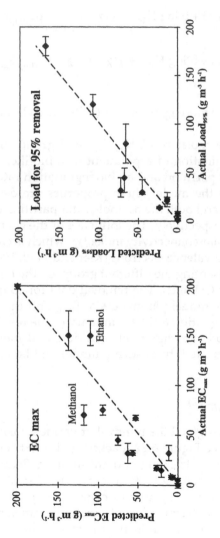

Figure 6.21 Observed and modeled values for the maximum elimination capacity and the load at which 95% removal occurs. The dashed line (slope 1) represents the perfect agreement between QSAR and experimental data.

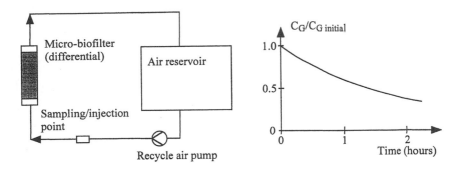

Figure 6.22 Schematic of the experimental setup and of the concentration pattern obtained in the micro-batch biofilter by Govind et al. (1997).

$$\mu = \sum_{groups} n_{group} \cdot \beta_\mu^{group} \qquad (6.16)$$

$$K_S = \sum_{groups} n_{group} \cdot \alpha_K^{group} \qquad (6.17)$$

The results showed that the group contribution for μ and K_S correctly predicted the behavior of the test compounds in a lab-scale, non-differential biofilter packed with a similar medium. As for the previous QSAR models, such an approach could eventually be used to estimate biofiltration kinetics with no prior biofiltration data.

6.5 Summary

In this chapter, selected biofilter models were described and their features exposed. By no means could this chapter review all of the numerous models that have been published recently. However, this selection is representative of the approaches generally used to model and simulate biofiltration experiments.

A summary of the main characteristics of each of the reviewed models is listed in Table 6.1 (pages 114–115). Interestingly, the general applicability of biofilter models for design is still questionable. The values of model parameters remain uncertain, and the sometimes complex equations are difficult to solve. Design is still often based on laboratory-scale biofilter test, extrapolation from published material, or field experimentation.

chapter seven

Design of biofilters

7.1 Introduction

Thirty years of experience have accumulated in the design and construction of modern biofilters (Figure 7.1). Through experimentation and experience, systems have evolved dramatically. Currently, because no two waste off-gases have the same contaminant, concentration, flow rate, temperature, and humidity, the ability to design an effective biofilter involves a combination of fundamental biofilter knowledge, practical experience, and bench- and pilot-scale testing. This design strategy produces flexible, cost-effective systems without the mistakes of past design and construction efforts. This chapter provides an in-depth discussion of biofilter design and construction, as well as associated problems. It describes a general protocol that should be followed in designing and implementing any biofilter system (Figure 7.2). Preliminary investigations of the waste stream and both bench- and pilot-scale experiments produce the necessary details to assess the effectiveness of the technology. Along with modeling, results from such experiments can be incorporated into sizing and designing full-scale systems. For full-scale implementation, various systems and components are discussed, and recommendations are provided for a successful scale-up of the technology. The topics discussed include reactor configuration, vessel construction, filter bed medium, air distribution systems, waste gas preprocessing controls, moisture control systems, and computer control and analytical systems. Based on these design choices, a cost-estimate analysis for a full-scale system is also presented.

Figure 7.1   Modern biofilter design (700 m³ filter bed volume) treating air from a solvents/plastic industry. (Photograph courtesy of Monsanto Enviro-Chem Systems; St. Louis, MO.)

7.2   Experimental protocol for assessing biofilter technology

For a particular site, a clear and concise definition of the air emissions problem should be established (see Figure 7.2). The characteristics unique to the site include the contaminants and concentrations to be treated, process flow rate, pressure, relative humidity, temperature, particulate concentration, regulatory removal efficiency required, and budget. Some of this information may be unknown or inaccurate, so preliminary testing may be warranted to assess the characteristics of the feed waste gas. This knowledge should guide a literature review on other biofilter research to assess the potential effectiveness of the technology at the particular site. Modeling may also be useful in predicting potential biofilter performance. Because past biofilter research may have been done at sites with different characteristics, bench and pilot testing may be required. This testing will determine whether biofiltration is the appropriate air pollution control technology for the specific waste stream. Additionally, from such testing, results should demonstrate the type of medium, microbes, and reactor conditions crucial to the success of the technology before the system is brought to full scale.

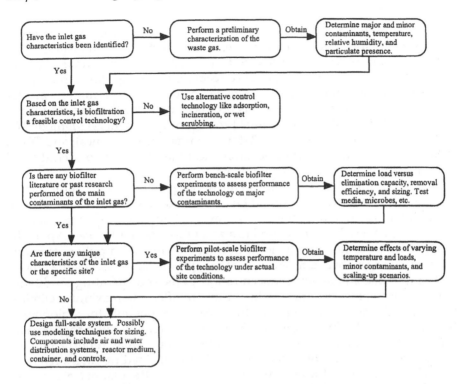

Figure 7.2 Decision-making process in planning and designing an effective full-scale biofiltration system.

7.2.1 *Preliminary testing and assessment of technology*

Preliminary testing should involve an in-depth analysis of the waste gas stream. Parameters to be tested include major and minor contaminant concentrations, loading rates, humidity, temperature, and particulate concentrations. The minimum, average, and maximum values should be measured over an extended period of time. A normal operating day may suffice, but plant (or site) operation should be assessed over a 2-week period to define periodic variations. This is especially true for sites with variable loads, such as those from batch processes. For example, a 34,000 m³ h⁻¹ (20,000 cfm) biofilter was operated over 90% of the time in an underloaded condition compared with the design loading specification. This resulted in an overall control efficiency lower than the design efficiency (Gilmore and Briggs, 1997). It is critical to the success of the biofilter to identify the average loading rate from the site process before implementation of a biofilter system. Not only is it more expensive to construct and operate an overdesigned system, but the necessary regulatory control efficiency requirements may not be met for a significantly underloaded biofilter. Underdesigning a system can be equally harmful, so a balance based on the average load must be achieved.

To comprehend accurately and adequately the composition of the waste gas stream before any bench-scale testing is performed, managers, field engineers, and personnel familiar with the process stream should be consulted. Such personnel will be knowledgeable about the specific input chemicals of the process and the flow rates to be treated. However, minor constituents may not be known if numerous products are formed or the feed is not well defined. Typical Environmental Protection Agency (EPA) Methods TO-11 (aldehydes and ketones), TO-12 (total non-methane organics), and TO-14 (chlorinated and non-chlorinated aromatics) may be used to identify and quantify major and minor contaminants.

With a complete identification of the waste gas stream and conditions, an extensive literature review should be performed to assess the feasibility of biofiltration. A literature review may reveal elimination capacities and removal efficiencies that can be used to size a full-scale biofilter (see Appendix B). However, the literature review information will only supply elimination capacities and removal efficiencies that are specific for certain media, microbial consortia, contaminants, concentration levels, and operating conditions. Performance values reported in literature are often the maximum elimination capacities with low removal efficiencies. It is possible for a biofilter to have large overall elimination capacity with a low removal efficiency because the mass loading rate is high. For example, literature may report a biofilter to remove 20 g m^{-3} h^{-1} of a contaminant, but if the loading rate were 100 g m^{-3} h^{-1}, the removal efficiency would only be 20%. Regulations are usually based on removal efficiency and not mass removed, so when assessing biofilter technology for a new site based on past research, caution is justified. A literature review may also reveal a past or current operating biofilter with similar waste gas characteristics and operating regimes. Such a system design may simply be duplicated for a new site if all system differences can be identified and their effects on overall performance are minimal.

Modeling techniques may be utilized to design and size a biofilter. Estimates of elimination capacities can be made using numerous models (see Chapter 6). If parameters such as biofilm thickness, interfacial surface area, and biokinetic and chemical constants for each contaminant and the microbiology are well understood, modeling may effectively predict elimination capacity and removal efficiency for a specific contaminant or a group of contaminants in a biofilter and aid in sizing the system. However, many of the biofilter properties are poorly known and may only be estimated. Estimated parameters may be based on past research that does not replicate the new potential site. These unknowns will reduce the predictive capability of the model. It is possible to overestimate many of these unknown variables and drastically oversize a biofilter system. Additional research will be helpful in confirming the data.

7.2.2   Bench-scale testing

Bench-scale testing is effective if it is performed on a waste stream that closely resembles the actual process stream. Such testing should be used primarily to demonstrate elimination capacity and removal efficiency for a particular contaminant or a group of contaminants as a function of mass loading. Using bench-scale experiments, the loading conditions (with the volume remaining constant) can be altered in one of two ways. The flow can be increased, increasing the loading rate and decreasing the residence time. The increase in flow will decrease the time allowed for the contaminant to diffuse into the biofilm and degrade and may limit the removal of the contaminant. Alternatively, the load can be increased by increasing the concentration while allowing the retention time to remain the same. In this case, the higher VOC concentrations provide a stronger driving force for diffusion into the biofilm and possibly faster biodegradation kinetics. However, a threshold exists above which concentrations may be toxic to the microbes. Even so, in most industrial biofilter applications, contaminant concentrations remain well below this toxicity threshold. For full-scale biofilter systems, this suggests efforts should be made to capture contaminants effectively so that concentrated streams can be more efficiently treated.

When operated at similar extreme loading rates, high flow/low concentration and low flow/high concentration conditions will produce different elimination capacities because of the different driving forces of removal. If concentrations of the actual waste stream are not expected to change over time, then mass loading vs. elimination capacity experiments should be conducted by only increasing the flow rate. If concentrations are expected to change (as in soil vapor extraction off-gases), both flow rate and concentration should be altered to study their effects on performance. Testing under various conditions will provide the data needed for successful modeling.

Bench-scale experiments will produce a typical elimination capacity vs. load curve (see Figure 1.7, page 20). This curve describes two performance scenarios based on either high removal efficiency or high elimination capacity. If high removal efficiency is desired, the biofilter should be operated at lower mass loads where the slope of the line is unity. As the loading on the system increases, mass loading and elimination capacity diverge at a point defined as the critical elimination capacity. If a high elimination capacity (a large amount of VOC degradation) is required, the biofilter should operate above the critical elimination capacity where the slope of the graph falls to zero. The VOC removal will be maximized (maximum elimination capacity), but removal efficiency will be lower. Commonly, regulatory permits are defined in terms of removal efficiency. The elimination capacity vs. load curve will define the obtainable elimination capacity within the constraint that removal efficiency will meet or exceed regulatory requirements (95% removal, 85% removal, etc.).

Performance study results will provide critical information for sizing a larger system. By knowing the highest bench-scale elimination capacity that meets the regulatory criteria, the design engineer can scale up to treat larger flow rates with a larger volume filter bed. For example, methyl ethyl ketone (MEK) removal was 95% in a biofilter at a MEK load of 68 g m^{-3} h^{-1} (MEK

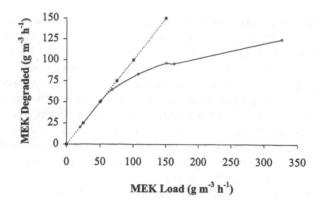

Figure 7.3 Elimination capacity vs. loading rate for a biofilter treating methyl ethyl ketone emissions. Flow rate is 0.4 m³ h⁻¹ and EBRT is 40 s. (Adapted from Deshusses, M.A., Biodegradation of Mixtures of Ketone Vapours in Biofilters for the Treatment of Waste Air, Ph.D. thesis, Swiss Federal Institute of Technology, Zurich, 1994.)

inlet concentration of 0.76 g m⁻³, flow of 0.4 m³ h⁻¹, EBRT = 40 s = 0.011 h), giving an elimination capacity of 64.5 g m⁻³ h⁻¹ (Deshusses, 1994; see Figure 7.3). Based on treating a similar concentration of MEK at a larger flow rate, the minimal volume of filter material required to give a mass loading that produces a 95% removal efficiency can be calculated (see Section 1.5; Equations 1.1 or 1.6). If contaminant concentrations are different at the bench level than at the pilot- or full-scale level, care must be taken to ensure the residence time for the pilot or full-scale system is similar to the residence time for the bench-scale system on which the calculations are based; otherwise, a safety factor will be needed in the volume of filter bed. From the above example, if the actual MEK inlet concentration for the full-scale system is 0.76 g m⁻³ (similar to the bench conditions) and the amount of air to be treated is 50,000 m³ h⁻¹, 95% removal can be achieved with a minimal bed size:

$$V_f \text{ (at 95\% removal)} = EBRT \times Q = 0.011 \text{ (h)} \times 50,000 \text{ (m}^3\text{h}^{-1}) = 550 \text{ m}^3 \quad \textbf{(7.1)}$$

or

$$V_f \text{ (at 95\% removal)} = \frac{C_{Gi} \times Q}{\text{Mass loading}} = \qquad\qquad\qquad \textbf{(7.2)}$$

$$= \frac{0.76 \text{ (g m}^{-3}) \times 50,000 \text{ (m}^3\text{h}^{-1})}{68 \text{ (g m}^{-3}\text{h}^{-1})} = 550 \text{ m}^3$$

If greater removal percentages are required by regulations, then more reactor volume will be required to lower the mass loading (volumetric) and improve removal efficiency across the bed.

Figure 7.4 Typical bench-scale biofilter. Numerous media are being tested for their ability to treat wastewater off-gases. (Photograph courtesy of T.S. Webster.)

Biofilter performance may be enhanced at the bench level by optimizing various operating parameters of the system. Different bed materials can be tested to assess their ability to host a thriving microbial population (Figure 7.4). Performance based on filter bed temperature, pH, and moisture and nutrient content variations can be analyzed, and different microbial inocula can be tested. Specific microbes that effectively treat the contaminants can be identified and cultured for later inoculation into the pilot- and full-scale systems. If the test duration allows for it, any toxicity problems caused by the major gas contaminants will be seen. All of the operating parameters can be varied at different loading rates (for both changing flows and concentrations). Because the possibilities are numerous, a good biofilter bench-scale project should begin with an extensive literature review to avoid duplication of previous efforts. Timely, high-quality, bench-scale experiments can determine the viability of the technology before more expensive pilot tests are performed.

7.2.3 Pilot-scale testing

Bench-scale experiments can effectively demonstrate the ability of a biofilter to treat the major air contaminants of a particular waste stream under favorable conditions (optimal temperature, relative humidity, etc.). However, results from bench biofilter studies under laboratory conditions sometimes

Figure 7.5 A typical pilot-scale biofilter reactor treating actual soil vapor extraction off-gases at a jet fuel spill. This system was designed by The Reynolds Group (Tustin, CA).

fail to predict performance seen at the pilot- or full-scale level (Lipski et al., 1997). This may be attributed to the presence of minor constituents and particulates, temperature and relative humidity variations, and varying inlet loads at the pilot- or full-scale level. Thus, pilot studies are best performed at the site with a side stream of the actual waste gas (Figure 7.5). As with the bench studies, elimination capacity vs. load experiments should be performed. During such testing, specific problems that may be encountered with the actual process off-gas can be identified and possibly resolved. Monitoring over an extended period of time will establish average and spike contaminant concentrations and their effects on the operation of the system. Information concerning the formation of any harmful degradation by-products from the minor constituents in the waste gas can also be evaluated. Research at the pilot scale, as well as at the bench scale, may also demonstrate possible problems with scale-up of the biofilter. With increasing flow rate, diffusion limitations in the biofilm may develop and reduce performance of the system at the full scale (Ottengraf et al., 1983). Therefore, extensive experimentation at both bench and pilot scales is recommended to establish an effective air velocity and pollutant loading that provide optimal removal so similar conditions can be achieved at the full scale.

7.3 Design of full-scale biofilters

Combining information obtained from an effective literature review, modeling, and bench- and pilot-scale experiments prepares the engineer to effectively design a full-scale biofilter system. These design tools demonstrate the effectiveness of a technology for a particular site while providing necessary sizing information. However, numerous considerations must be addressed at the full scale when developing the individual components of the system (Figure 7.6). The bioreactor configuration, reactor vessel, filter medium, air distribution system, pretreatment system, moisture control system, and computer control and analytical system should be designed to provide the most effective biofilter possible within the technical and financial constraints of the project.

7.3.1 Reactor configuration

The simplest biofilter design is the open bed system (Figure 7.7). With this design, compost or soil media is placed in or on a supporting structure (covering the air distribution system) and exposed to the atmosphere at the top surface (Allen and Yang, 1991; Leson and Winer, 1991). The open bed is commonly 1 to 1.5 m in depth, may be either above or below grade, and is a bottom-loaded system. Open-bed systems are ideal for applications where space is not a constraint and are often used as low cost/low performance biofilters for odor control. They may be affected by weather conditions that can reduce operational performance. Heavy rainfall may saturate the media, creating air flow problems. An inexpensive cover placed over an open biofilter may be advisable in wet regions. During winter months, freezing conditions may require inlet air heating to maximize reactor performance.

An enclosed biofilter is more complicated in its design. It may be designed in many shapes, including cylindrical and rectangular, and provides for better control of operational parameters such as temperature, humidity, water content, nutrients, and pH (Figure 7.8). The bed is generally 1 to 1.5 m in depth, is composed of organic or inorganic medium, is above grade and closed to the atmosphere, and can be operated as either a top- or bottom-loaded system. Weather has less effect on enclosed systems. This allows a more favorable microbial environment to be maintained. Increased control results in optimal performance and generally reduces space requirements.

Open and enclosed biofilter systems may be operated as single-level or multiple-level beds, alone, in parallel, or in series. For both single- and multi-

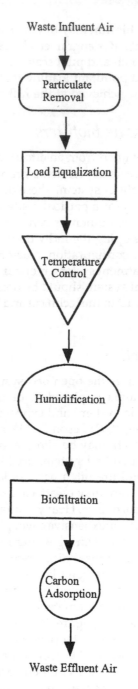

Figure 7.6 Possible treatment train for a biofiltration system. If load equalization is unnecessary, particulate removal, temperature control, and humidification are often accomplished in one step. The carbon adsorption step is optional.

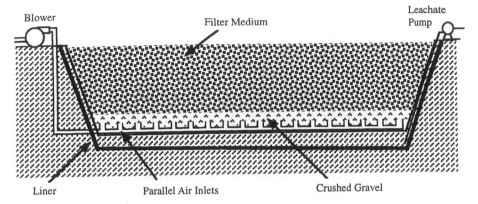

Figure 7.7 An open biofilter bed, below grade, bottom-loaded (forced draft), with parallel air inlets and crushed gravel to distribute the air evenly through the bed.

level biofilters using organic material as a bed medium, filter bed depth per level should not exceed 1 to 1.5 m to avoid compaction. If a deeper bed is required, additional structured support material should be added to the medium. Depending on the contaminants to be treated and the air flow rate required, a simple single-level bed biofilter may be sufficient (Figures 7.7 and

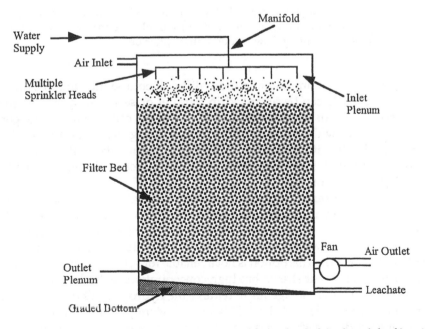

Figure 7.8 An enclosed biofilter bed, above grade, top-loaded (induced draft), with an inlet and outlet plenum designed to distribute air through the system homogeneously.

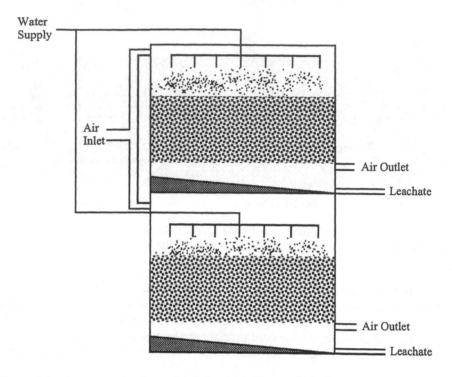

Figure 7.9 An example of a multiple-level biofilter with beds operated in parallel. This configuration allows the isolation of individual beds for maintenance or shut down without interrupting flow through other filter beds.

7.8). If space is not limiting and more difficult compounds are to be treated, single-level beds can be arranged in parallel or series. If space is limiting, multiple-level beds operated in a parallel or series can provide for longer gas retention times while using a smaller footprint (Figures 7.9 and 7.10). Finally, for larger loads with difficult to degrade contaminants, multiple-level beds may be operated in parallel or series with other multiple-level beds for complete treatment.

Parallel and series configurations each have advantages and drawbacks. Systems operated in parallel offer the flexibility to isolate individual beds for maintenance without completely shutting the biofilter down. A biofilter operated in series allows for the treatment of individual contaminants through different filter beds. This option is not available for systems operated in parallel. For example, if an air stream has high concentrations of hydrogen sulfide and VOCs, it may be advantageous to remove the hydrogen sulfide in one biofilter operated at low pH and the VOCs in an ensuing biofilter at a neutral pH. However, a drawback to systems operated in series is the large head loss generated against the air delivery system. Regardless of the type of configuration, as the bed numbers increase, more infrastructure will be required.

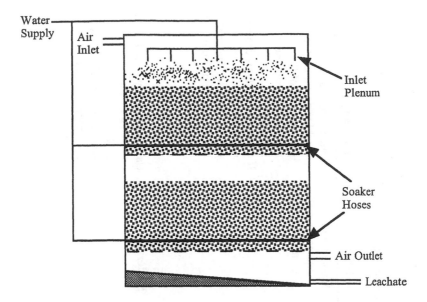

Figure 7.10 An example of a multiple-level biofilter with beds operated in series. If desired, this configuration allows each individual bed to treat different contaminants.

7.3.2   Biofilter vessel construction

For open and enclosed biofilters, suitable vessel construction materials vary with the particular application. The reactor vessels may be constructed from a variety of different materials, including polyvinyl chloride (PVC) or polyethylene plastic, fiberglass-reinforced polyester, cinder block, concrete, and stainless, carbon-coated, or galvanized steel (Table 7.1). The materials for construction will depend on the particular configuration (open, closed, single-level, multi-level, above or below grade), sophistication desired, and the waste gas characteristics. Material considerations include weight, workability, and cost. Additionally, the choice and amount of material for construction will require careful consideration by the design engineer so that the chemical and physical characteristics of the incoming waste stream do not damage the structural integrity of the vessel. For example, at a site using three modular biofilters in series, an aggressive mixture of VOCs, temperature, and pressure encountered from a glycol dehydrator process severely stressed reactor hatch seals and seams. On numerous occasions, all gaskets had to be replaced and sealed with silicone to repair leakage problems. Painted surfaces (epoxy solvent resistant paint) quickly failed as well (Stewart and Kamarthi, 1997).

Open-bed systems placed below grade can be designed with or without a constructed reactor vessel. If the soil quality is firm, excavated pits may not require structured siding at all. The filter material can simply be placed

Table 7.1 Biofilter Reactor Vessel Construction Materials

Materials	Durability	Ability to work with	Relative cost	Special notes
Carbon steel	Rugged, heavy material Will not crack or bend easily Requires coating that may include galvanizing, powder-coating, metalizing, painting, etc. Coatings may become scratched or contain imperfections, leading to corrosion of steel	Requires welding work, but only after coating is removed	Moderate to expensive cost for infrastructure	Coated carbon steel may be used for portions of large, permanent biofilter units to provide structural integrity If coated properly, it can be used as a grating to prevent biofilter material from entering plenum
Cinder block	Moderate weight material Must be coated to prevent moisture penetration Chips and imperfections in coating may cause moisture to enter and reduce the cinder block strength Susceptible to failure under tensile loads	Comes in premolded blocks that can be fitted together Difficult to cut or shape once cinder block has set	Moderate	Used extensively in full-scale systems treating over 3400 m^3 h^{-1} Precast cinder blocks allow for repetitive, easy installation for large systems

Material	Weight/properties	Handling/construction	Cost	Applications
Concrete	Moderate to heavy in weight; Must be coated to prevent moisture penetration; Chips and imperfections in coating may cause moisture to enter and reduce the concrete strength; Reinforced concrete performs better under tensile loads	Can be poured into molding to meet design specifications; Difficult to cut or shape once cement has set	Moderate	Used extensively in full-scale systems treating over 3400 m^3 h^{-1}; Precast concrete modules allow for repetitive, easier installation for large systems; a crane is usually needed to move modules
Fiberglass epoxy	Moderate weight material; Susceptible to cracking and chipping if handled improperly but fairly resistant to weathering	Moderate difficulty in cutting and shaping after resin has set; Produces fine particulates when cut; Requires more effort in adding fittings	Moderate	Used primarily in smaller biofilter applications (<8500 m^3 h^{-1}); however, can prove to be suitable for larger systems, as seen for scrubber construction at wastewater treatment plants
Galvanized steel	Rugged, heavy material; Will not crack or bend easily; Susceptible to scratches and scrapes, reducing corrosive resistance effectiveness	Requires welding work, but only after galvanizing coat is removed	Moderate cost for infrastructure; Inexpensive for ducting	Galvanized steel may be used for portions of large, permanent biofilter units to provide structural integrity; Generally will be used as grating to prevent biofilter material from entering the plenum

Table 7.1 (continued)

Materials	Durability	Ability to work with	Relative cost	Special notes
High-density polyethylene or polyvinyl chloride	Lightweight; resistant to corrosion. May break down with UV exposure, weakness in plastic causing cracks and. Susceptible to collapse if large internal vacuum pressures are encountered	Extremely easy to cut and contour for the addition of fittings and piping attachments	Inexpensive to moderate for reactor container. Increases exponentially in cost with increasing pipe diameter	Generally used for smaller pre-molded biofilter units and humidification towers
Stainless steel	Rugged, heavy material. Will not crack or bend easily. Extremely durable; very resistant to weathering and effects of corrosion	Requires welding work	Very expensive	May be used as grating to prevent media from entering plenum. Generally too expensive to use on very large systems. Welds may still be susceptible to corrosion
Wood	Lightweight. Requires weatherproof coating. A tendency to crack or chip easily. May rot if not coated properly	Easy to cut and shape. Will not provide leak-proof construction	Inexpensive	Effective to maintain structural integrity. Not advisable to be used where corrosive gases may come in contact

inside a pit on top of the air distribution system (van Lith et al., 1997). Such systems may require a liner and leachate collection system. The liner material should be chemically and biologically resistant, with a large water-holding capacity. Materials such as high-density polyethylene are effective and have been used in landfill systems for over 30 years. A leachate collection system should be installed at the base of the excavated pit. Excavating the pit with a slight gradient at the base and a drain at the lower end will allow for leachate collection. This water can be either pumped out of the system and treated or recirculated over the bed. Gravel can be placed directly over the leachate collection and air distribution system. This gravel layer prevents filter bed material from clogging these systems. A coarse geotextile fabric may be spread on the top of the gravel layer to prevent clogging of the air holes by particles of medium.

For above-grade, open-bed systems or for below-grade, open-bed systems without firm soil quality, sidewall construction will be necessary. These sidewalls may be constructed of materials such as concrete, coated steel, or wood. Baffles may be added along these walls to create better air distribution. A leachate collection system and a gravel layer will also be required.

Construction materials for enclosed systems will depend on the size, shape, location, and use of the biofilter. For small biofilters, materials should be utilized that are light in weight and resistant to weathering. It is possible to purchase premolded containers made of plastic, polyethylene, or PVC materials (Figure 7.11). These containers are light and fairly inexpensive, requiring minimal labor to set up. If the handling of larger loads is required, these smaller systems may be set up in parallel (Figure 7.12). For larger systems, more rugged materials are advantageous. Larger systems generally will be constructed from a combination of materials that add structural strength to the system while minimizing cost. Modules of steel-reinforced concrete or epoxy-coated steel may be precast and brought on site individually (Figure 7.13). These modules can then be added into the existing infrastructure using a crane. If designed properly, lighter biofilter systems may be placed on the roof of a building, providing an advantage to companies with space limitations (Figure 7.14).

7.3.3 Filter bed medium

Careful consideration must be given to the choice of filter bed material so that its life expectancy is optimized and performance is maintained (Chapter 3). As time progresses, the aging packing material may cause an increase in pressure drop because of microbial deterioration and compaction. Additionally, diminishing nutrient supply or pH buffer depletion may cause a decline in microbiological activity. Such problems may only be resolved through material replacement, usually occurring every 3 to 7 years (Figure 7.15).

It is important to choose a medium with optimal chemical, physical, and microbiological characteristics. Such parameters to consider when choosing

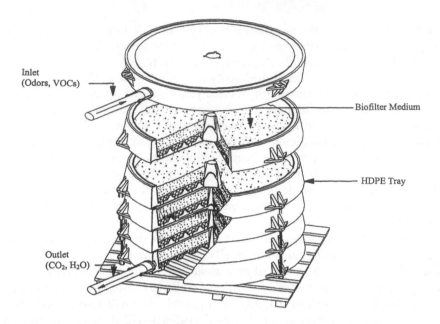

Inlet
(Odors, VOCs)

Biofilter Medium

HDPE Tray

Outlet
(CO$_2$, H$_2$O)

Figure 7.11   Schematic of a lightweight commercial biofilter. This system incorporates up to five stacked, high-density polyethylene trays filled with biofilter medium to treat contaminated air in series. (Drawing courtesy of AMETEK-Rotron Biofiltration Products; Saugerties, NY.)

Figure 7.12   Where space is not a limitation, smaller biofilter systems may be placed in parallel to treat larger loads. (Photograph courtesy of AMETEK-Rotron Biofiltration Products; Saugerties, NY.)

Figure 7.13 Pre-cast concrete modules being brought on site enabling easy construction of the system. (Photograph courtesy of Envirogen, Inc.; Lawrenceville, NJ.)

Figure 7.14 Where space may be limited on the ground, lighter biofiltration systems have been manufactured that can be placed on the roof of the facility. (Photograph courtesy of Monsanto Enviro-Chem Systems; St. Louis, MO.)

Figure 7.15 Medium replacement for a full-scale biofilter system. The procedure is labor intensive and time consuming. (Photograph courtesy of Monsanto Enviro-Chem Systems; St. Louis, MO.)

an effective filter material include surface area, porosity, water-retention capabilities, nutrient supply, microbiological activity, and life-cycle costs. While organic material, such as compost and soil, will have an indigenous microbial population and nutrient supply, they will have to be added to inorganic materials. Organic media will need to be sieved properly to eliminate small particles (less than 4 mm in diameter; see Allen and Yang, 1991). This will prevent increases in backpressure and the potential for clogging of the reactor inlet or outlet air orifices (Allen and Phatak, 1993; Corsi and Seed, 1994). Inorganic media may be quite uniform, so that sieving is not needed. Inorganic material, such as granular activated carbon (GAC), polystyrene spheres, and ceramics have been shown to be effective filter materials with good adsorptive characteristics and porosity (Liu et al., 1994; Medina et al., 1995b; van Groenestijn et al., 1995). Such materials provide an additional buffering capacity for spikes or shockloads of contaminants; however, the initial capital costs are much higher than those for organic materials. If acid-forming contaminants are present in the waste gas, both medium types will require the addition of alkaline buffers, such as calcium carbonate, to neutralize acid intermediates generated by microorganisms. The reduction in this buffering capacity will be a function of the load on the system. It will eventually diminish, requiring material removal and caustic recharging. Frequent bed removal will increase operating costs of the system. The best

Figure 7.16 Air distribution system transporting air to a biofilter system. (Photograph courtesy of Monsanto Enviro-Chem Systems; St. Louis, MO.)

choice of filter medium will ultimately be a function of its life expectancy and cost-effectiveness. Combinations of organic and inorganic materials may be advisable to obtain the benefits of both material types.

7.3.4 Air distribution systems

The primary objective in designing an air distribution system is to transfer contaminated air from the process outlet to the inlet of the biofilter for homogeneous distribution through the filter bed. The gas must first be collected from the generating process. For point source emissions, a large collection hood may be placed near the process outlet source or the contaminated air may be delivered directly to the biofilter through ventilation ducting (Figure 7.16). For fugitive emissions, a gas collection system will require larger capturing hoods and a substantial ventilation system. The hoods and ducting may be constructed of PVC, fiberglass-reinforced epoxy resin, stainless steel (14 or 16 gauge), galvanized steel (14 or 16 gauge), or sheet metal (Grades 22–30), among others. The choice of material will depend on many physical and chemical parameters of the contaminated air: the pollutant type, contaminant concentration level, temperature, humidity, and intended flow rates. The costs associated with each type of material will also require assessment. In general, streams consisting of corrosive contaminants at high concentrations, streams with high temperatures and low humidity, streams with

large flow rates, and biofilters operating in fluctuating environmental conditions (drastic changes in ambient temperature) will require more chemically
resistant, rigid material. For example, a waste gas from a hardboard mill
operation containing aldehydes, alcohols, ketones, fatty acids, and aromatics
was treated using a full-scale biofilter. The galvanized steel inlet ducting was
badly corroded within 3 months of operation because of the corrosive nature
of the inlet air (Allen and Van Til, 1996). Because the characteristics of every
waste gas are different, general recommendations are difficult. Each particular
application must be addressed by the design engineer. The use of material
resistance tables will provide a starting point. However, pilot experiments on
various construction materials may be warranted to assess durability and
cost-effectiveness prior to full-scale implementation.

A biofilter may be top or bottom loaded using a forced draft system
(pressure) or an induced draft system (vacuum; see Figures 7.7 and 7.8). An
open biofilter requires a forced draft to distribute the contaminated air
upward through the filter material. For enclosed systems, either system can
be used. Forced draft systems require that contaminated air flow through the
blower, and there is a greater likelihood of corrosion damage. Additionally,
if leaks occur, untreated contaminated air may be forced out through these
leaks. An induced draft system ensures that only treated air contacts the
blower. If leaks are present in the system, clean air will be drawn into the
system. Top-loaded enclosed systems may be advantageous because irrigation water is applied where the drying is most severe (Table 7.2). However,
bottom-loaded systems may work well for air streams requiring additional
humidification before entering the filter bed because water irrigation is
applied at the bottom of the reactor (through soaker hoses).

The inlet of a biofilter should distribute contaminated air evenly through
the filter bed. For open systems, slotted pipes made of chemically and biologically resistant material (such as PVC) can be installed at the base of the
reactor and covered with crushed rocks or gravel (Figure 7.7). If spaced
properly, the slotted pipes evenly distribute the air through the filter material. Openings in the pipes can be placed farther apart where gas pressures
are high to make flow more uniform. The rock cover prevents filter bed
material from entering the slotted pipes and clogging the distribution system. Some biofilter systems have used hollow precast concrete blocks or
corrosion-resistant grating at the base of the filter for both clogging control
and accessibility for heavy equipment. These systems require more capital
investment and have not always prevented clogging. For enclosed systems,
an air distribution system can be designed so that air enters a plenum
(Figures 7.8 and 7.17). For both top- and bottom-loaded systems, the gas flow
velocity and plenum geometry must be considered so that the air is homogeneously distributed through the bed.

Control of the air flow through the distribution system and biofilter will
require proper valves. For systems with small inlet lines (less than 30 cm in
diameter), ball valves may be installed in line to control flow rate. For larger

Table 7.2 Important Considerations in Deciding Whether a Biofilter
Should Be Top- or Bottom-Loaded

Air distribution through a biofilter	Important considerations
Top-loaded	Water irrigation and biological activity occur predominantly in similar areas at the top of the bed.
	Eliminates the possibility of water hold-up on filter material.
	By-products such as acids will percolate down through the filter bed, possibly damaging the structure of the material.
	Particulates may clog inlet area, where microbial oxidation will be diminished.
	Particulates cannot be flushed readily from the biofilter without possibly affecting the remainder of the bed.
	Hazardous chemicals will be at the surface of the material, making maintenance difficult. The system will require shutdown and ventilation prior to entry.
Bottom-loaded	Area of surface irrigation and majority of biological oxidation occur in different regions of bed.
	There may be water hold-up near areas of filter where air channeling occurs. Anaerobic zones may appear.
	By-products can easily be removed from the system through soaker hoses located near the inlet of the bed. Damage to the bed is minimal.
	Particulates can be flushed out of system without media damage.
	Surface of bed can be easily examined since contaminant concentrations should be low at the outlet.

inlet lines, dampers may be more appropriate for flow control. Both valves and dampers will be necessary at the inlet and outlet of the biofilter if bed isolation is required. If large, possibly toxic concentrations are expected in the waste gas, dilution valves should be installed in the air distribution system at the inlet of the biofilter. The materials of the flow control devices should be compatible with the waste gas contaminants to prevent breakdown or corrosion of components such as o-ring seals on valves.

The choice of a particular blower or fan will primarily depend on the desired flow rate through the biofilter. Whether forced or induced draft, air-feed systems are rated for certain flow rates at a maximum head. This head will be a function of a vacuum or pressure head across a humidifier, air distribution inlet and outlet ducting length and diameter, filter bed material and porosity, biomass build-up, and any other in-line devices (particulate drop-out pod, pre-screen filters, etc.). Underestimating head loss will cause inadequate sizing of a blower. In one case, this caused low-flow problems for

Figure 7.17 The base structure and plenum design for a bottom-loaded, dual-inlet biofilter. (Photograph courtesy of Envirogen, Inc.; Lawrenceville, NJ.)

a full-scale biofilter when pressure losses across the bed exceeded design estimates (Allen and Van Til, 1996). This head-loss calculation must be made to ensure that the blower will operate at the desired flow rate. Head-loss calculation is sometimes difficult because it requires estimation of the pressure drop increase as biomass forms and the filter bed compacts over time. A rough estimate of this increasing head can be made, but, if time permits, long-term media experiments may be warranted to better define the actual losses before a blower is purchased. Oversizing the blower should be avoided for economic reasons. Larger blowers incur higher capital costs and may require installation of additional flow regulating valves. Variable speed blowers are more expensive, but allow maintenance of a constant flow as pressure drop increases.

The blower or fan material should be chosen carefully. Because the air will be treated as it is pulled through a filter bed, an induced draft fan will generally contact lower concentrations of the contaminated air. Thus, compatibility with the contaminants is less critical; however, a forced draft blower will contact untreated air, particulates, high temperatures, and moisture. A blower will rapidly lose efficiency if incompatibility of the blade material allows formation of precipitates or pitting of the blades. This can cause long downtimes for a biofilter, and repair or replacement of the blades can be difficult and expensive.

7.3.5 Waste gas preprocessing

Many biofilter systems have experienced poor reactor performance and continuous system operational problems because of insufficient knowledge of the inlet waste gas characteristics. Identification of the inlet gas contaminant species, concentrations, and loading conditions is essential. Additionally, information on the particulate concentration, inlet waste gas temperature, and relative humidity of the inlet stream must also be obtained. The gas must be preconditioned to minimize its effect on the microbial ecosystem while maximizing the performance of the reactor.

7.3.5.1 Contaminant species and concentrations

Knowing all the contaminant species and the concentrations being produced may help avoid performance problems with a biofilter. Many organics and inorganics that appear in the air phase will not be harmful to the microorganisms if their concentrations are low. Although toxic threshold air concentrations depend on the Henry's Law coefficients and the specific toxicity of the contaminant, gas contaminant concentrations exceeding 5 g m^{-3} of organics or 3 g m^{-3} of inorganics are often toxic, and an investigation of toxicity is advisable if they are anticipated. If occasional spikes of concentration are expected in the waste gas, a carbon adsorption load equalizer may be installed upstream of the biofilter. This carbon will serve as a buffer, desorbing during lower concentration periods such as weekends and late nights (Ottengraf et al., 1986; Weber and Hartmans, 1995).

7.3.5.2 Particulate emissions

Biofilter reactors are designed to be self-regenerating biological systems. However, biofilters will also act as effective filters for particulate emissions. Particulates that enter a biofilter with the waste gas may easily clog the medium. It is imperative that particulate loading be avoided or treated before entering the biofilter. The production of particulates will be a function of the specific process that is creating the waste gas. Plant personnel and detailed analysis can identify the physical and chemical nature of the particulates. Plant or facility engineers will have a thorough understanding of their process and should be familiar with its particulate generation. Environmental Protection Agency (EPA) Methods 5 and 202 can be used to assess PM10 (particulate matter less than 10 μm in diameter) in the air stream. Failure to perform such extensive testing or simply being unaware of them has caused many operational problems for biofilters.

Associated problems with unfiltered particulates entering biofilter systems include, among others, the deposition of high-molecular-weight compounds (oils, fats, greases) and chemical reactions within humidification chambers which create foaming problems. A biofiltration company reported on many such problems (Standefer and Willingham, 1996). In treating a formaldehyde-laden waste gas with a pilot-scale unit, engineers discovered that cooling the waste gas through a packed humidifier caused condensation of formaldehyde resin particulates that clogged the medium. This system required a pretreatment filter to remove these condensed resins before the air entered the biofilter reactor. In another biofilter system treating off-gases from a wood-products operation, condensables increased the suspended and dissolved solids level of the humidifier recirculation water, resulting in foaming. Such problems require ongoing maintenance and can be avoided only through careful scrutiny of the waste gas characteristics. Depending on the particulate size and nature, simple pre-treatment filters, air humidification scrubbers, high-energy venturi scrubbers, or wet electrostatic precipitators will remove particulates. Many of these processes can be used simultaneously to adjust the inlet waste gas temperature and relative humidity levels.

7.3.5.3   Temperature

In general, biofilters tend to operate effectively in the mesophilic temperature range (20 to 45°C) where the most diverse microbial communities thrive. For this reason, it is important to supply the waste gas at a temperature as close to this mesophilic range as possible. This temperature will have a large effect on microbial degradation kinetics, as well as the moisture content of the biofilter medium (see Section 7.3.5.4). Rates of microbial reactions and diffusion will increase with rising temperature. Microbial activity will increase by a factor of 2 for each 10°C rise up to approximately 35°C; however, phase transfer and sorption will become less favorable with rising temperatures. For industrial processes that produce waste gases at temperatures far exceeding the mesophilic range, gases must be cooled. This cooling may be performed through adiabatic cooling during humidification or by heat exchangers using a cooling fluid. Conversely, in a few examples of low loading biofilters, effective removal was seen even at low temperatures (5 to 16°C; see Section 9.6). For higher loads, the systems may be oversized to provide longer contact times to compensate for the possible lower level of biological activity. Gas preheating will allow reduction in the size of the biofilter and may be accomplished through direct steam injection, forced heating, or through packed towers with heated recirculating water. However, if heating or cooling of the inlet waste gas is required, biofiltration may not be economical. Additionally, cooling of the warm, humid air generates wastewater. For example, 50,000 m^3 h^{-1} saturated air at 40°C cooled to 25°C generates 1400 L h^{-1} of water. This could be a problem for zero discharge facilities.

7.3.5.4   Humidification for moisture control

An essential component for biofilter success is the control of moisture content in the filter bed. A dry filter medium will cause a severe decrease in microbial activity and poor treatment of waste gases. The drying of filter bed material may also cause bed shrinkage and channeling of air. Depending on the hydrophobicity of the medium, these effects will be difficult to reverse without agitating or removing the filter bed material (Paul and Nisi, 1996).

In the situation where the filter medium becomes too wet, numerous problems may also occur: clogging of pore spaces, development of anaerobic zones, increase in pressure drop, and leaching of nutrients. The relative humidity, temperature, and residence time of the waste gas entering a biofilter will cause either evaporation or condensation of water between the filter bed and the passing air. Control of filter bed moisture content can be accomplished through the use of a humidification chamber upstream of the biofilter and direct irrigation of the filter bed (see Section 7.3.6).

The need for a humidification chamber before the biofilter reactor will depend on the inlet waste gas temperature and relative humidity and on the filter bed temperature. If the inlet waste gas and the filter bed are at equilibrium, neither evaporation or condensation will occur. This is rarely the case because microbial oxidation is an exothermic reaction. The inlet gas will not be in equilibrium with the solid/liquid phase, and assuming an adiabatic system, the heat generation will cause a bed temperature increase and create a decrease in the humidity of the gas. Moisture loss from the medium will occur on a continuous basis, and the filter bed will eventually dry. To prevent this problem, a humidification chamber is usually set up prior to or after the blower system to maintain a 95 to 100% relative humidity waste gas at the appropriate bed temperature. If the chamber is installed upstream of the blower, gas temperature increase caused by the blower will decrease the relative gas humidity.

Humidification devices maximize the air/liquid interface so that the gas relative humidity achieved is near 100%, and the temperature is maintained within the appropriate biofilter range. This means the rate of drying of the biofilter medium will be kept minimal and further moisture control can be accomplished using direct irrigation. Some humidification systems may also be used to remove particulates. These systems require either a separate filtration chamber to remove the particulates from a recirculating water phase or a continuous purge of wastewater to maintain the concentration of suspended solids at a reasonable level. Typical humidification chambers include steam injection systems, spray chambers, venturi scrubbers, or packed bed towers. For smaller biofilters (<8500 m^3 h^{-1}), spray chambers have proven effective (Figure 7.18). For larger flow biofilter systems, packed bed towers are more frequently used (Figure 7.19). The choice of humidification device will be a function of both space limitations and operating costs.

Typical problems of humidification systems are clogged packing and water carry-over to the filter bed. Past designs have used cross-flow packed bed humidifiers, but some have become clogged with biomass in as little as 3 months (Mildenberger and van Lith, 1996). Clogging requires increased maintenance and operating costs. Water carry-over may overwet the inlet portions of the biofilter. This condition can cause excessive filter bed compaction and nutrient leaching. Compaction may require that the filter bed be tilled periodically to avoid the development of anaerobic zones. Nutrient leaching may require periodic nutrient addition to the biofilter to avoid

Figure 7.18 Typical spray humidification chamber (white cylindrical tank on left) used for smaller biofilter applications to maintain a saturated inlet vapor. (Photograph courtesy of Envirogen, Inc.; Lawrenceville, NJ.)

performance declines. These costly maintenance scenarios can be avoided if the humidification chamber is designed appropriately and, if needed, a demister is installed.

7.3.6 Water irrigation for moisture control

Humidification alone may not succeed in maintaining the moisture content of the media bed at optimal values. Hence, an irrigation system is essential for a successful biofilter. Any effective irrigation system requires that filter bed moisture be monitored periodically. Water loss in a biofilter may be determined manually, or automatically using a programmable logic controller (PLC; see Section 7.3.7.1). Manual methods include (1) measuring temperature increases across the filter bed and calculating water loss through the use of standard psychrometric charts, and (2) measuring weight loss of the filter bed through the use of load cells. Automatic methods that may be preprogrammed through the PLC include (1) conventional moisture probes and (2) conductivity corrected moisture probes. From such measurements, either manual or automatic irrigation may be performed. An automatic irrigation system will be timer controlled to activate for a specified amount of time. An automatic system may also be more sophisticated, controlled by

Figure 7.19 Pack bed tower (black) saturates the contaminated air before being processed through numerous biofilter beds operated in parallel. (Photograph courtesy of AMETEK-Rotron Biofiltration Products; Saugerties, NY.)

a PLC that will use information from moisture probes to maintain the moisture content of the media. The system complexity, as well as capital cost, increases when an automatic system is used; however, if it is designed correctly, labor costs may be dramatically reduced.

7.3.6.1 Irrigation systems

Irrigation of the filter medium may involve spraying the surface with a water hose or the use of an overbed sprinkler system, soaker hoses, or pressure-compensated irrigation hoses (Table 7.3). Manual spraying should be done carefully to ensure uniform application. The use of a flow-control device on the hose will aid in determining the volume of water delivered to the bed. For a sprinkler system, a main water line will deliver the water through a manifold to individual pressure compensated sprinkler heads (Figure 7.8). Operated in parallel, these heads can spray water evenly over the entire surface of the bed. Soaker hoses will release water into the filter medium through tiny orifices in the hose wall. A modified version of the soaker hose is the pressure compensated irrigation hose (Figure 7.20). This hose has evenly spaced discharge points, each with tiny pressure-compensating valves. Both hose types, depending on the air flow direction, may be set up at the top, middle, or bottom of the filter bed. For top-loaded systems, hoses should be

Table 7.3   Advantages and Disadvantages Associated
with the Installation and Operation of
Various Irrigation Systems in Biofilters

Irrigation system	Advantages	Disadvantages
Water hose/manual	No need for system set up Controlled amount of water added to top of bed	Increased operator interaction Possibility of uneven watering of filter bed surface
Sprinkler system/ automatic	Uniform distribution of water on surface of filter bed Controlled amount of water added to filter bed	More complex irrigation Infrastructure required on biofilter Possibility of sprinkler heads clogging
Soaker hoses/ automatic	Slower release of controlled amount of water in bed ensures homogeneous filter bed wetting Hoses placed near inlet of bottom-loaded systems allows for additional humidification of entering air Possibility to easily remove condensables and particulates from the media bed near the inlet Hoses are fairly inexpensive and similar to agricultural type hoses	More complex irrigation Infrastructure required on biofilter Differential pressures throughout hose length vary the volume of water delivered to the filter bed Clogging of orifices on the soaker hoses is possible, preventing any water from entering the biofilter bed Difficult to perform maintenance on hoses in bed without removing the media
Pressure-compensated soaker hoses/automatic	Same as regular soaker hoses but pressure-compensated valves on each orifice let operator strictly regulate flow volume of water into the bed	Same as regular soaker hoses Pressure-compensating valves may become clogged

Note: All systems, with the exception of the manual water hose, can be operated either manually or automatically.

Figure 7.20 Typical soaker hoses placed at the bottom of a biofilter to provide for additional humidification of air as it enters the bed; a bottom-loaded pilot system is shown with filter material to be placed over the hose. (Photograph courtesy of Envirogen, Inc.; Lawrenceville, NJ.)

placed at the top and middle of the bed so that gravity moves the liquid downward. For bottom-loaded systems, hoses should be positioned at the base of the filter bed (above the plenum and grate) and at the top of the bed. The bottom hose provides humidification of the incoming air, while the top hose is used for conventional irrigation. The choice of the irrigation system will be application specific. Sophisticated systems may require pressure-compensated irrigation hoses throughout the bed controlled by a sophisticated PLC, while simple systems may just require that a sprinkler system be occasionally turned on manually.

7.3.6.2   *Operational problems associated with irrigation systems*

Water irrigation systems are susceptible to operating problems, but these can be avoided if clearly understood. Water nozzle or sprinkler choice will be critical. Numerous past biofilter pilot- and full-scale systems have been plagued by inadequate nozzle design (Mildenberger and van Lith, 1996). Clogging of the nozzles appears to be the key problem that limits their performance. In general, industrial type, chemically resistant nozzles with large orifices that can provide adequate water distribution at low pressure should be used. The sprinkler system should be designed for easy access in the event that the sprinklers require cleaning. For soaker hoses, the material of the hose should be chemically resistant. Soaker hoses are similar to standard garden hoses and often do not withstand the rigors of the biofilter environment. Splitting and cracking are possible. Additionally, discharge points on the hoses are not pressure compensated, causing different flow volumes to be released from the discharge points along the length of the hose. These variations make it difficult to ensure that irrigation is homogeneous irrigation. This problem can be avoided through the use of pressure-compensated irrigation hoses. However, the tiny pressure-compensating valves may also be occasionally clogged. The design of an effective water irrigation system requires awareness of the various types of problems associated with water irrigation equipment.

7.3.7 Computer control and analytical systems

Computer control and analytical systems required for a biofilter will depend on the reactor size, type of monitoring required, and funds available. The larger the biofilter, the greater the need for extensive control and monitoring. Basic analytical equipment will be needed to assess the performance of a biofilter, but costs will vary depending on the types of monitoring conducted. Regulatory permits will dictate the required control and analysis of any biofilter system. However, minimizing control and analysis instrumentation to reduce capital costs may prove detrimental to the overall success of the system. With the overall objective of minimizing capital and operating costs, a balance must be established between the required monitoring equipment necessary to meet the regulations and any additional analytical equipment to ensure adequate biofilter performance.

7.3.7.1 Computer control systems

Control and monitoring of a biofilter system may be performed manually or by a computer. Through manual operations, a blower can be turned on and off, water and nutrients can be sprayed on the bed, and physical and chemical parameters can be measured using portable instruments. Manual operation is expensive if the system is significantly large or in a remote location. For larger systems that require more extensive control and monitoring, computer control is advisable. Programmable logic controller computers are increasingly integrated into biofilter design as more advanced and effective systems are constructed. These computers operate instruments that measure biofilter parameters such as filter bed moisture content, pressure drop, bed weight, air flow rate, temperature, relative humidity, and inlet and outlet concentrations. The PLC collects, saves, and visually presents all data so that performance status of the biofilter system can be assessed quickly. The PLC can also use a feedback-control program to document and resolve any abnormalities in system operation. For example, the PLC may turn on water sprinklers when moisture content is low or may shut the system down if filter bed pressure drop is excessive. This type of computer logic system minimizes the need for human interaction and provides for better control of system conditions. It may also prevent the damage of system components (water pumps, blowers, etc.) if biofilter malfunctions occur and are not recognized. The system may also include a modem for off-site data access and for paging maintenance personnel during an alarm condition. This is a necessity when the system is set up in an isolated area. Such advanced systems are expensive; however, if designed properly, the systems will substantially reduce the labor costs for a biofilter system.

7.3.7.2 Analytical systems

Various operating parameters can be measured using commercially available instruments (see Chapter 8 for further details). For the measurement of gas-phase concentrations, continuous gas monitoring equipment is best, but expensive. Gas chromatography (GC) is generally the simplest method for vapor analysis. A flame ionization detector (FID) can be used to measure total and specific hydrocarbon concentrations. A photoionization detector (PID) is effective in measuring organic and inorganic compound concentrations. Chlorinated compound analysis can be accomplished using a direct electrolytic conductivity detector (DELCD) or an electron capture detector (ECD). For odor measurements, an olfactometer may be used. For all gas samples, analysis may be performed on-line or grab samples may be taken in non-reactive sampling bags or stainless steel canisters and injected into a GC. For on-line sampling, care must be taken to prevent moisture from condensing in the sample lines. This may be accomplished by warming the lines with heating tape, by using appropriate membrane filters, or by periodically backflushing the lines with ambient air.

Additional parameters such as bed and gas temperature, flow rate, pressure drop, bed weight, and relative humidity should be determined to assess the operation of a biofilter. Temperature can be effectively measured by thermocouples. Measurements from an orifice plate with a differential pressure gauge can be used to calculate air flow. Pressure drop can be measured by manometers or differential pressure meters. Bed weight can be analyzed by placing the reactor or its compartments on load cells. Humidity probes can measure relative humidity. The extent of monitoring, beyond what is required by regulatory permits, will depend on available funds and the types of data desired.

7.3.7.3 Electrical requirements

Electrical service available at plants and sites varies — 110-volt alternating current (VAC), 240-VAC, 480-VAC, single-phase, three-phase, etc.; therefore, the design engineer must be aware of the service available when designing the mechanical, computer control, and analytical systems. For large installations, the electrical requirements of the biofilter will generally dictate what type of power will be brought onto the site. If a different source of power is required for the system than the already established source at the particular site, the addition of the source by the local power company will be a small fraction of the total system cost. However, for smaller biofilter systems, where allowable expenses are much smaller, the installation of a new service at the site may be cost prohibitive. The engineer should consult with site electricians, managers, and electrical drawings before sizing and purchasing a blower, pump, gas analyzer, or other electrical equipment. This will ensure the availability of a specific electricity type, proper wiring, and suitable circuit breakers. In the U.S., all industrial plants carry 110-VAC electricity. However, some plants carry both 240- and 480-VAC (or similar variations in voltage such as 208/220 VAC and 440/460 VAC), while others utilize only 480-VAC because of the cost savings. Additionally, industrial plants may have only single-phase power, three-phase power, or both. Available wiring gauge — 12 American wire gauge (AWG), 14 AWG, etc. — and circuit breakers may not be sufficient to handle high-amperage loads. During design and construction, attention to such small details may avoid the need for major adjustments (such as the addition of an expensive transformer). It will also allow the proper electrical equipment to be installed well before the reactor is delivered or constructed, saving much labor and time in the beginning stages of reactor installation.

7.4   Costs and economic considerations

The decision to use a biofilter will undoubtedly be a function of its cost. In general, a regulatory biofilter performance goal, usually expressed as percent removal, is established. In designing the system components of a biofilter, the design engineer's goal is to meet a performance level while minimizing capital (investment) and operating costs. Capital costs include all system equipment and labor to build and install the reactor, including costs attributed to depreciation and interest. Operating costs arise from energy consumption, water consumption and disposal, monitoring, maintenance, and medium replacement. In comparing past biofilter applications, some general cost trends can be established. However, making more specific capital and operating cost assessments is difficult because of the differences in waste gases, performance requirements, and system designs. Using a simple model, an attempt at optimizing biofilter design to minimize cost has been performed with some success (Gerrard, 1997). Unfortunately, this approach to economizing a biofilter system still requires numerous approximations. In this section, general capital and operational costs are discussed. A detailed breakdown of capital and operating costs for a hypothetical system design is also presented. Actual costs for various biofilter designs are available in Chapter 9.

7.4.1   Capital (investment) costs

Capital costs have been reported for a number of biofilter applications (Togna et al., 1994; Austin et al., 1995; Boyette et al., 1995; Leson and Smith, 1997). For a given size, open-bed systems with simple designs will generally be less expensive than enclosed-bed systems with more control options. As systems increase in volume, the rate of cost increase declines. This is partly because the cost of materials will be reduced through large-quantity discount purchases. Also, the amount of effort required to design and construct a larger system will not be significantly greater. Costs per unit volume vary greatly. Reactor costs for small designs (100 m³; 3500 ft³) have been estimated at $1000 to $3500 per m³ of filter bed ($28 to $99 per ft³ of filter bed). Larger designs (3000 m³; 110,000 ft³) become more cost effective at $300 to $1000 per m³ of filter bed ($9 to $40 per ft³ of filter bed; see van Lith et al., 1997).

Investment costs are also often calculated on the basis of cost per unit volume of air treated ($ m⁻³ h⁻¹, $ cfm⁻¹). Costs per unit volume of air decline as the residence time decreases because filter bed volume decreases (if flow is constant). Investment costs range between $5 and $150 per m³ h⁻¹ capacity ($9 to $255 cfm⁻¹), with averages in the range of $7 to $35 per m³ h⁻¹ capacity ($11 to $60 cfm⁻¹) (Jol and Dragt, 1988; Fouhy, 1992; Deshusses, 1994; van Lith et al., 1997). These cost estimates are also a function of the degradability of the contaminants. For compounds that are easily degradable, costs are significantly lower (Figure 7.21). The higher cost estimates are based on more advanced, computer-controlled, enclosed systems treating compounds difficult to degrade (Figure 7.22).

7.4.2 Operating costs

Operating costs are primarily a function of energy consumption, water consumption and disposal, monitoring requirements, maintenance, and media replacement. All of these operating costs vary from system to system. However, generalized costs have been reported to range from $0.1 to $3 per 1000 m³ ($0.1 to $3 per 35,000 ft³) of waste air treated (Jol and Dragt, 1988; Eitner, 1990; Fouhy, 1992; Deshusses, 1994). The degree of expense attributed to each cost varies based on the design of the system and the air stream being treated. Like the investment costs, operational costs are a function of the type of contaminant treated (Figures 7.21 and 7.22). Air streams with easily biodegradable contaminants may require a smaller bed volume, minimizing the electrical demand, water consumption, and monitoring and maintenance required to operate effectively. Larger systems will be necessary to handle more recalcitrant contaminants. Such larger systems will have larger overall pressure drops, increasing electricity demand. In addition, these systems will require more water to maintain the moisture content of the larger filter bed. Finally, monitoring and maintenance will be more time consuming and may require additional personnel to effectively operate the system.

7.4.2.1 Energy consumption

Energy consumption by a biofilter is a substantial part of operating costs. Electrical equipment such as water pumps, analytical equipment, and

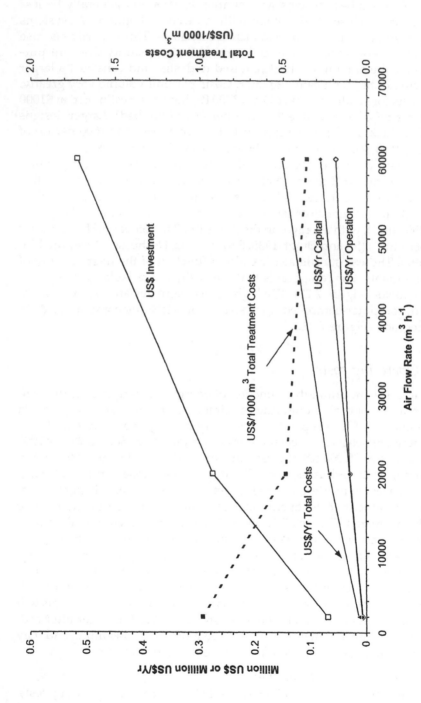

Figure 7.21 Biofiltration capital and operating costs (~20%) vs. flow rate for systems treating compounds that are easy to degrade. Based on 8% interest rate over 15 years. Assumes 7000 hours of operation per year; 1 Fr = $0.69 US. (From BUWAL, *Ablufreinigung mit Biofiltern and Biowaschern*, BUWAL Schriftenreihe Umwelt Nr. 204, BUWAL, Bern, Switzerland, 1993. With permission.)

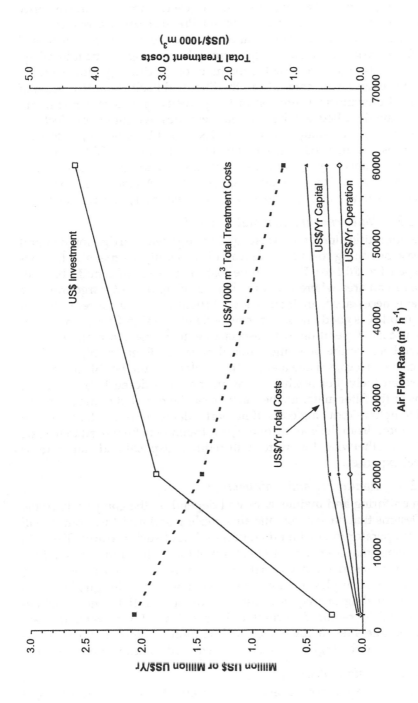

Figure 7.22 Biofiltration capital and operating costs (~20%) vs. flow rate for systems treating compounds that are difficult to degrade. Based on 8% interest rate over 15 years. Assumes 7000 hours of operation per year; 1 Fr = $0.69 US. (From BUWAL, *Ablufreinigung mit Biofiltern and Biowaschern*, BUWAL Schriftenreihe Umwelt Nr. 204, BUWAL, Bern, Switzerland, 1993. With permission.)

computer equipment will all require some electricity; however, the majority of electrical costs arise from the blower or fan. Pressure drops across the inlet air ducting, humidification chamber, and filter bed media create an increased demand for electricity. Pressure drops across the ducting can be minimized by limiting bends and increasing the diameter of the ducting. However, the majority of pressure drop comes from the filter bed itself. A large portion of this electrical demand is a function of the porosity, moisture content, and structure of the filter bed. As the material degrades over time, the electrical demand increases. Generally, compost and soil biofilters will operate with pressure drops less than 500 Pa (5 cm of water column) and 1500 Pa (15 cm of water column), respectively (Bohn, 1996). Systems with inorganic medium will operate with pressure drops similar to compost beds. An increase in pressure drop will correspond to a substantial increase in electricity consumption.

7.4.2.2   Water consumption and disposal

Extensive humidification of the waste gas or large water irrigation demand will increase operating costs. Potable water is generally inexpensive to moderately expensive ($0.1 to 0.7 m^{-3}). These costs may be significant if a biofilter is operated in an area where water is scarce or if significant purge from the humidifier is necessary to maintain a low particulate matter concentration. If water availability is good, the water costs will not be a significant contributor to operating costs. Operating costs associated with leachate formation should be minimal because the quantities should be small. Even for high BOD or acidic leachates, a local wastewater treatment plant should be able to treat the water inexpensively. For leachates that are not considered harmful to the organisms or the filter material, the leachate can be recycled to the top of the filter bed to help in reducing costs. If no particulates are present, condensed water generated from the cooling of the gas stream may be reused to irrigate the filter bed; otherwise, this water will require disposal and can increase operational expenses.

7.4.2.3   Monitoring and maintenance

Routine monitoring and maintenance will depend on the complexity of the system. Designs that have automatic air sampling and moisture control will require less operator interaction than those without such features. This will reduce the labor requirements for operation of the system but will require a larger capital investment. A cost-savings analysis is needed to assess the benefits of a more complex monitoring system. For poorly designed systems where air channeling or frequent upsets occur, substantially more maintenance time will be required to resolve the problems. For systems that are located in remote sites, maintenance will be more expensive. Such operational costs justify the higher capital costs required to automate a system.

7.4.2.4   Media replacement

Organic media are replaced every 3 to 7 years. As the medium ages, it contributes to pressure drop and increased electricity demand. Medium replacement is a demanding task, and the time for removal and replacement, labor required, and medium cost will vary based on the size of the reactor and the medium utilized. Systems designed for easy medium access will produce substantial cost savings in terms of labor requirements. Additionally, specialized inorganic media may have extended bed life and perform more reliably but will require a higher initial capital investment.

7.4.3 Cost estimate example for a full-scale biofilter

A cost estimate for construction and operation of a full-scale biofilter requires numerous assumptions that may be different depending on the location of the system, the characteristics of the waste air, and regulatory requirements. The example developed here is based on a system designed to treat 17,000 m³ h⁻¹ (10,000 scfm) of air containing low concentrations of odors with no particulates. Bench- and pilot-scale results demonstrated an effective elimination capacity at a 70-s empty bed retention time (Webster et al., 1995). From such results (assuming a 20% safety factor), an open-bed, compost-based biofilter with dimensions of $20 \times 20 \times 1$ m ($66 \times 66 \times 3.3$ ft) is being constructed. The biofilter has perforated piping embedded at the base of the system to provide contaminated air (see Figure 7.7). Moisture content within the biofilter is controlled by humidification of the air before entering the reactor, as well as by direct irrigation through soaker hoses at the top and base of the packed bed.

7.4.4 Full-scale biofilter cost example

Assume Q = 17,000 m³ h⁻¹ (10,000 cfm); low odor concentrations; 95% contaminant removal at an EBRT = 70 seconds; open-bed dimensions = $20 \times 20 \times 1$ m; continuous operation throughout the year (Webster et al., 1995).

7.4.4.1 Capital costs

A. Initial site preparation costs
   1. Volume of media bed = Q × EBRT
      $V = 17{,}000$ (m³ h⁻¹) × 70 (s) × 1 (h)/3600 (s)
      $V = 330$ m³
   2. Assuming a 20% safety factor:
      V safety = V × 1.2 = 330 (m³) × 1.2 = 396 m³ or L × W × H of 20 (m) × 20 (m) × 1.0 (m)
   3. Assume another 0.305 m for the gravel bed height:
      Volume of gravel is 0.305 (m) × 20 (m) × 20 (m) = 122 m³
   4. Volume to be excavated = media volume + gravel volume:
      Volume to be excavated = 396 (m³) + 122 (m³) = 518 m³
   5. Excavation (includes equipment, labor, overhead/profit) and off-site disposal (assume non-hazardous and 10-mile roundtrip) = $7.1 m⁻³
   6. Total cost for site preparation = $7.1 (m⁻³) × 518 (m³) = **$3700**

B. Medium costs
    1. Assuming \$37 m^{-3} for compost:
        Medium cost = \$37 (m^{-3}) × 396 (m^3) = \$14,700
    2. Assume \$33 m^{-3} for gravel:
        Gravel cost = \$33 (m^{-3}) × 122 (m^3) = \$4030
    3. Installation of the medium and gravel:
        Equipment rental = \$0.7 m^{-3}
        Labor = \$1.1 m^{-3}
        Profit/overhead = \$0.7 m^{-3}
        Total for equipment, labor, and profit/overhead = \$2.5 m^{-3}
        Total installation of medium costs = \$2.5 (m^{-3}) × 518 (m^3) = \$1300
    4. Total medium costs = \$14,700 + \$4030 + \$1300 = **\$20,000**

C. Equipment
    1. One blower: standard 10-hp at **\$2500** each
    2. One packed bed tower humidifier: **\$2000** each

D. Piping (10% of capital costs)
E. Electrical (4% of capital costs)
F. Equipment installation (4% of capital costs)
G. Engineering design (10% of capital costs)
H. Liner required
    1. A_{liner} = A + (L) (H) (2) + (W) (H) (2) = 396 (m^2) + 20 (m) × 1.305 (m) × 2 + 20 (m) × 1.305 (m) × 2 = 500 m^2
    2. Liner cost for 3-mil PE (including stabilization) = \$22 m^{-2}, so total cost for liner = \$22 (m^{-2}) × 500 m^2 = **\$11,000**

I. Mobilization and demobilization of construction crew and equipment (**\$5000** lump sum)
J. Miscellaneous (**\$5000** lump sum for permitting, spare parts, etc.)

Total capital cost: X = A + B + C1 + C2 + D + E + F + G + H + I + J

- X = \$3700 + \$20,000 + \$2500 + \$2000 + 0.10 (X) + 0.04 (X) + 0.04 (X) + 0.12 (X) + \$11,000 + \$5000 +\$5000
- Total capital cost: X = **\$70,300**
- Piping cost = 10% of capital cost = **\$7030**
- Electrical cost = 4% of capital cost = **\$2810**
- Equipment installation = 4% of capital cost = **\$2810**
- Engineering design = 12% of capital cost = **\$8440**

7.4.4.2 Annual operation and maintenance costs

A. Electricity (one blower and one water humidification pump)
    1. Annual blower electrical consumption:

$Consumption_{blower} = 10$ (hp) $\times$ 0.75 (kW hp^{-1}) $\times$ 365 (days yr^{-1}) $\times$ 24 (h day^{-1}) = 65,700 kWh

2. Annual pump electrical consumption:
   $Consumption_{pump} = 0.25$ (hp) $\times$ 0.75 (kW hp^{-1}) $\times$ 365 (days yr^{-1}) $\times$ 24 (h day^{-1}) = 1650 kWh
3. Total electrical consumption = 65,700 (kWh) + 1650 (kWh) = 67,350 kWh
4. Total electrical cost = 67,350 (kWh) $\times$ \$0.08 (kWh)$^{-1}$ = **\$5400** yr^{-1}

B. Water consumption
   1. Assuming inlet air from process stream is 50% saturated, it is estimated that 36 m^3 wk^{-1} (9500 gal wk^{-1}) will be used. Assume cost of water is \$0.70 m^{-3}.
   2. Water consumption = 36 m^3 wk^{-1} $\times$ 52 wk yr^{-1} $\times$ \$0.70 m^{-3} = **\$1300** yr^{-1}

C. Labor (1 hr day^{-1} at \$20 h^{-1}):
   Annual cost of labor = \$20 h^{-1} $\times$ 365 h yr^{-1} = **\$7300** yr^{-1}
D. Overhead (25% of labor costs):
   Annual cost of overhead = \$7300 $\times$ 0.25 = **\$1800** yr^{-1}

Total yearly operating costs = A + B + C + D

- \$5400 + \$1300 + \$7300 + \$1800 = **\$15,800** yr^{-1}

7.4.4.3 Medium replacement costs (every 5 years)

A. Medium removal cost (similar to 7.4.1, step A)
   1. Excavation, transportation, and disposal = \$7.1 m^{-3}
   2. Volume to be excavated = 396 m^3
   3. Total cost for medium removal = 396 (m^3) $\times$ \$7.1 m^{-3} = **\$2800**

B. New medium addition (similar to 7.4.4.1, step B)
   1. Assuming new medium costs of \$37 m^{-3} for compost:
      Medium cost = \$37 (m^{-3}) $\times$ 396 (m^3) = **\$14,700**
   2. Assume medium installation costs of \$2.5 m^{-3} (includes equipment, labor, and profit/overhead).
   3. Total installation of medium costs = \$2.5 (m^{-3}) $\times$ 396 (m^3) = **\$1000**

Total medium replacement cost (every 5 years) = A + B1 + B3

- \$2800 + \$14,700 + \$1000 = **\$18,500**

7.4.4.4 Annualized costs

A. Capital costs
   1. Assume an 8% interest rate over 15 years.

2. Annualized capital cost = $70,300 × (A/P, 8%, 15) = $70,300 × (0.1168) = **$8200** yr^{-1}

Note: A/P = annual payment/present value of money.

B. Operating costs
1. Assume a 3.5% rate of inflation over a 15-year time frame:

$$\text{Total cost over 15 years} = \sum_1^n \$15,800 \,(F/P, 3.5\%, 15) = \$315,000$$

Note: F/P = future value of money/present value of money; n = number of years.

2. Annualized operating costs (over 15 years) = **$21,000** yr^{-1}

C. Medium replacement costs
1. Assume a 3.5% rate of inflation and an 8% interest rate over 15 years. The medium will be changed after 5 and 10 years only.
2. Annualized medium replacement costs = $18,500 × (F/P, 3.5%, 5) × (P/F, 8%, 5) × (A/P, 8%, 15) + $18,500 × (F/P, 3.5%, 10) × (P/F, 8%, 10) × (A/P, 8%, 15) = $18,500 × (1.19) × (0.6806) × (0.1168) + $18,500 × (1.41) × (0.4632) × (0.1168) = **$3200** yr^{-1}

Annualized costs = A + B + C

- $8200 + $21,000 + $3200 = **$32,400** yr^{-1} (±20%)

7.4.4.5   Other cost estimates

A. Investment costs per unit volume
1. Capital costs/volume of biofilter bed
2. $70,300/396 (m^3) = $178 m^{-3} of biofilter bed

B. Investment costs per flow rate
1. Capital costs/flow rate:
2. $70,300/17,000 (m^3 h^{-1}) = $4.1 m^{-3} h^{-1}

C. Operating costs per volume of air treated
1. Operating costs per 1000 m^3 of air treated
2. In one year, 150 million cubic meters of air will be treated at a flow rate of 17,000 m^3 h^{-1}
3. $15,800/1.5 × 10^5 (m^3) = $0.11 per 1000 m^3 of air treated

D. Annualized costs per volume of air treated
1. Annualized costs per 1000 m^3 of air treated
2. In one year, 150 million cubic meters of air will be treated at a flow rate of 17,000 m^3 h^{-1}
3. $32,400/1.5 × 10^5 (m^3) = $0.22 per 1000 m^3 of air treated

7.4.5 Summary

This cost estimation of a simple, in-ground biofilter demonstrates only some of the numerous possibilities and choices that must be considered before construction. This example demonstrates the relationship between design choices and their economic impact to the overall system costs. For instance, if the system is designed to be above ground and enclosed, costs for excavation, transportation, and disposal of soil could be eliminated; however, the cost of materials for the reactor vessel would be added. Also, a more complicated design requiring local government approval would be required (earthquake-safe, etc.). Adding a sophisticated control and monitoring system could easily increase the cost and complexity of the design. It would reduce operating costs but also increase capital costs. These are just a few of the hundreds of small details that need to be considered and weighed for their economic advantage. It is only through the understanding of these complex design choices and their economic impact on the overall annualized costs that a successful, cost-effective, full-scale biofilter can be designed, constructed, and operated.

7.5 Conclusions

The design of a biofiltration system requires a detailed understanding of site conditions, site limitations, system components, and costs. In general, most biofilter failures occur because of poor "problem definition" before design and construction of the system. A preliminary assessment of the particular site should be performed so that all site conditions are known. These conditions include waste gas composition and concentration, humidity, temperature, and particulate content, among others. In addition, ambient site conditions that could affect the system components should be noted. A thorough literature review, modeling, and bench- and pilot-scale tests will provide the knowledge for full-scale system design.

Design options at full scale are numerous. Critical system components that require careful consideration include the reactor configuration and vessel, filter bed medium, air distribution system, pretreatment system, irrigation system, and the control and analytical systems. The desired operational control over the system will dictate whether an open or enclosed bed system configuration is better. The choice of organic or inorganic media as filter bed material will depend on such factors as its ability to host a thriving microbial community, durability, and life-cycle costs. The type of air distribution through the filter bed, forced or induced draft, and the need for pretreatment will depend on the type of contaminants being treated. The water irrigation system can be simple in design with an operator occasionally watering the bed surface, or it can be extremely advanced through the use of pressure-compensated irrigation hoses operated by a PLC unit. Care should be exercised so that materials for all system components are compatible and rugged and will withstand both the harmful effects of the waste gas and the ambient site conditions. Ultimately, all the system components can be designed to be very simple or extremely complex. Biofiltration technology is generally chosen because of its cost-effectiveness compared with other air pollution control technologies. Some balance must be formulated between designing a system to be functional while making it sufficiently advanced and robust to operate under varying conditions.

chapter eight

Biofilter startup and monitoring

8.1 Startup

Design and construction of a biofilter system must be followed by rapid development of a vigorous microbial culture. "Startup" time is the acclimation time required to establish optimal biological removal. Depending on ambient and site conditions, this startup procedure may last for a few days to a few months. If a system is operated in a cold climate or the contaminant load is variable, startup times may be longer. They can be minimized if the medium is properly handled during loading and inoculation. Additionally, by slowly increasing the mass loading rate on the system at startup, toxic shock may be avoided as microbes acclimate gradually. There is also an opportunity to monitor the bed and to control water irrigation rates to optimize moisture content. Once a microbial population is adequately established on the medium, a monitoring program can be instituted to measure effectiveness and guide maintenance.

8.1.1 Material shipping and handling

Once a medium type has been identified for a full-scale system, proper shipping and handling will be required. Precautions should be taken so that the medium is kept moist, shielded from extreme temperatures, and protected from crushing and segregation. Its microbial population and structure should not be damaged. Sieving may be appropriate to remove smaller particles (<4 mm) and is commonly done by the supplier. An initial survey of compost media should be undertaken to remove any abnormally large pieces (twigs, large pieces of bark, etc.) which could substantially affect flow direction in the biofilter. In order to do this, a front-end loader may be required to turn the material.

The medium should be tilled and watered so it remains near optimal moisture content, and it should be stored away from direct sunlight to reduce evaporation. If too much water is applied, small particles may be washed through the medium, collecting where the flow slows and causing clogging. This can be prevented by watering the medium in stages, allowing sufficient time for soaking. If too little water is applied, compost may dry. The bed will shrink, and reestablishing optimal moisture content may be difficult. Even if excess water is added after drying, the hydrophobic nature of the material may cause the flow to channel. This will prevent rewetting of irregular volumes within the medium (Thompson et al., 1996).

The medium may tend to segregate by size during shipping and handling. This can be avoided by ensuring the medium is homogeneously moist, causing the smaller particles to adhere to the larger ones. The medium should be delivered to the site only after the system is completely prepared for filling, so that long storage periods are not necessary. If compost must be stored for short periods of time at the biofilter site (no more than 2 to 4 weeks), it should be tilled and watered to maintain the microbial population. It should also be stored at a temperature conducive to microbial growth (preferably from 5 to 45°C).

8.1.2   Loading of material

Preventing medium damage during the loading of the system is essential, as it is during shipping. If moisture content measurements are below the optimum, the material should be tilled and watered just prior to placement in the biofilter. The material should also be brought to the pH chosen for the application. Filter bed material should be loaded into the biofilter using a front-end loader rather than a conveyor belt system to reduce segregation by particle size and density. In order to prevent compaction problems, the total medium depth should not be so great that the material at the bottom is compacted by its weight. If depths greater than one meter are needed in order to reduce the biofilter footprint, compost should be fortified with other structurally resistant media (wood bark, foam pellets, etc.). Loading should be done in approximately 30-cm (1-ft) layers, with each layer raked, watered, nutrified, and, if necessary, inoculated. Baffles extending into the medium from the inside perimeter of the vessel may be installed between layers to prevent channeling where the medium meets the walls. Air flow should begin within 24 to 48 hours after inoculation and loading. This will prevent development of anaerobic conditions, which could be harmful to the microbial population. In addition, composting activity may resume and result in extreme temperatures (>60°C) which may damage the medium and alter the microflora (Deshusses, unpublished results). For this reason, also, the biofilter should go into operation soon after the compost is loaded.

8.1.2.1 Inoculation

Inoculation of the filter material may be required. Inorganic materials will certainly require a seed culture. Generally, composts will not require inoculum because a diverse microbial population will already be present; however, an inoculum may be added to minimize startup time. This inoculum may help maximize the system performance more quickly, especially if a more recalcitrant contaminant is being treated. The inoculum can be a general consortium with broad substrate specificity or a specialized culture with an affinity for a particular contaminant. Various inocula are available from wastewater treatment plants, universities, and commercial distributors. Wastewater treatment plants are a good source for activated sludge that can be used as a general consortium inoculum. Although the health risk is minimal, the use of activated sludge is being discouraged because of the possible presence of pathogens. Instead, consortia from the food processing industry that do not contain pathogens are being used. The inoculum can be mixed with the medium in a separate tank or simply sprayed on. In either case, the inoculum should be diluted with water to prevent clogging or clumping of the filter material. The inoculum must be homogeneously distributed throughout the filter material.

8.1.2.2 Nutrient and chemical addition

The addition of nutrients to the medium during the loading phase will depend on the filter medium utilized and the waste gas treated. Besides a carbon source, all microorganisms require small quantities of nutrients to metabolize the contaminant. The major nutrients required for this process are nitrogen, phosphorus, and potassium. Minor nutrients that may be required include iron, magnesium, calcium, zinc, manganese, and sulfur. If any nutrient is absent, the biodegradation process will stop. Compost media will generally have sufficient amounts of these nutrients, but additions may still be warranted. Commercial media are provided with sufficient nutrient concentrations, but nutrients may be added as aqueous solution of mineral salts or in a solid form suited for slow release. Morgenroth et al. (1996) demonstrated that increases in nitrogen concentrations caused immediate improvement in the performance of a compost reactor treating hexane. In all cases, inorganic media will require nutrient additions. Inexpensive agricultural fertilizers can be used for this purpose if the inoculum is a general consortium; however, a more specific formulation may be needed if pure cultures are being used.

Buffering chemicals may be added to the medium during loading if acid-producing gases are to be treated. Acid metabolites may degrade organic material, releasing small particles within the filter bed (Yang and Allen, 1994). These may clog the biofilter or air distribution system. The acids formed will also cause pH declines that can inhibit microbial activity. An initial charge of buffer materials in the medium will neutralize generated acids. Dolomite, crushed oyster shells, or similar additives can release carbonate as it

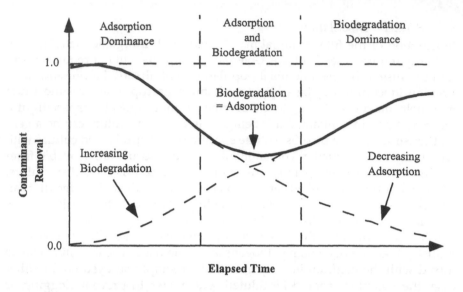

Figure 8.1 Adsorption and biodegradation during biofilter startup. (From Webster, T.S., Control of Air Emissions from Publicly Owned Treatment Works Using Biological Filtration, Ph.D. thesis, The University of Southern California, Los Angeles, 1996.)

is needed. Buffer should be added in amounts that will neutralize the acid generated during the expected life of the medium. The amount of buffer needed may be difficult to determine because of varying inlet loads and possible loss to leaching. As the buffer is exhausted, dilute concentrations of sodium bicarbonate or sodium hydroxide may be added to the biofilter through the irrigation system. However, concentrated solutions should be avoided, as they will harm the microbes on the medium. The frequency of additions will be a function of the contaminants treated and the mass loading rate. Alkalinity and pH should be monitored to determine when caustic will be required and at what rate.

8.1.3   Air flow rates and mass loading

Air flow rates should be 25 to 50% of the design flow rate for the first 3 days to 2 weeks. Pumping air at lower flow rates (but with normal concentrations) provides more time for the contaminants to diffuse and be oxidized in the biofilm developing on the packing material, while minimizing emissions to the environment. This lower loading should allow a thicker biofilm to readily develop in the first week or so. Initially, a large elimination capacity will be seen, resulting mostly from adsorption effects (Figure 8.1). Over a period of time, the length of which depends on the contaminant concentration and the adsorptive capacity of the medium, outlet concentrations will increase and elimination capacity will decline. Finally, the outlet concentrations will once again decline as microbiological activity on the filter material rises. The system should remain at this reduced loading condition until required removal is achieved.

After this initial period of low loading has nurtured an active biomass, the flow rate can be increased stepwise over short periods of time (1 to 2 days) until the design flow is achieved. After each increase in flow rate, an initial reduction in performance followed by an increase to a stabilized elimination capacity should occur. It is important to allow the system adequate time to stabilize at the new elimination capacity before increasing flow rates. As the flow rate is increased to the design flow rate, an elimination capacity should be established that meets the regulatory requirements of the operating permit.

8.1.4 Water irrigation

Approaches to the addition of water to filter beds during system startup depend on filter material characteristics and the design of the irrigation system. For all beds using either soaker hoses or sprinkler irrigation, water should be carefully applied. Too much water will cause leaching and possibly compaction, while too little water will cause bed drying. Initially, presuming the medium is already at optimal moisture content, water should be supplied frequently for short durations (2 to 5 minutes per hour of operation). Medium samples should be visually observed for compaction. Drains should be monitored for leachate, and if it is being produced, water addition rates should be decreased. Appropriate water addition rates can be calculated using the inlet air flow rate, temperature, relative humidity, bed temperature, and ambient conditions; however, these calculations assume ideal mixing conditions and may not be very accurate. Hence, water addition rates are generally based on experience and past system operation and should be adjusted as observations of biofilter conditions are made. Pilot-scale studies may help define optimal rates. Regardless of the initial water addition rate, the filter bed will require more water as the air flow rate and microbial activity increase.

More subtly, a water supply rate appropriate for the bed as a whole may not be good for each part of the bed if the sprinkling is not uniform. It is surprisingly difficult to design a system that adequately waters all of a surface without overwatering any of it. If the sprays from the sprinklers overlap, they may produce wet spots where both sprinklers apply water. If the sprays are reduced to avoid overlapping, dry spots may appear where they do not quite meet. If the spray strikes a wall, a substantial amount of water will be intercepted and will flow down the wall, either appearing as leachate or overwatering the adjacent medium.

For these reasons, immediately after the biofilter is built, and at intervals thereafter, some attempt should be made to monitor the uniformity of the watering. In the field, visual inspection of the surface of the bed may indicate whether there are obvious irregularities in the medium, such as compaction or deterioration, which are associated with overwatering. While no one has yet reported such a procedure, uniformity might be checked by distributing small open containers on the surface of the bed and comparing the amounts of water they collect, as is sometimes done for garden sprinklers.

8.1.5 Temperature

For most biofilters under ordinary conditions, the inlet gas temperature should be held between 20 and 45°C. As a result of biological oxidation, temperatures will increase as air flows through the bed. Outlet temperatures will be the highest. If outlet temperatures are greater than 50°C, inlet temperatures should be reduced. If the outlet temperature is less than 15°C, an increase in inlet temperature is recommended. As with water addition, incremental changes should be made to optimize the system during startup. Large temperature changes may damage the microbiological population and should be avoided.

8.2 Monitoring

For any biofilter to be successful, it is necessary to operate it so that the microbial culture is vigorous and healthy. The physical and chemical characteristics of the biofilter must also be controlled. If biofilter operating parameters are to be maintained at optimum values, they must be monitored. The remainder of this chapter reviews the parameters which have been monitored in past applications and describes the techniques for measuring each (Table 8.1). More sophisticated and innovative techniques are sometimes used in laboratory experiments. These are described here because it is likely that many of these will be applied in the field in the future. Suggestions are also made for techniques which have not yet been applied, but which would provide valuable information for biofilter operators. Some of the following material has previously appeared in the *Hazardous, Toxic and Radioactive Waste Practice Periodical* (1998) in the article, "Monitoring Biofilters Used for Air Pollution Control," by J.S. Devinny, and is reprinted with permission of the American Society of Civil Engineers.

8.3 Air load

The amount of air flowing into a biofilter and the contaminant concentration control the load, or mass of contaminant per unit time that the biofilter must treat. Air flows may vary with operating conditions in the source. Ventilation flows from work spaces, for example, may be higher during certain activities and much lower at night or on weekends. Over the longer term, slow compaction of the bed material and accumulation of biomass gradually increase

Table 8.1 Parameters for Biofilter Monitoring

Parameter	Method	Purpose
Air load		
Air flow (m³ h⁻¹ or scfm)	Input flow meters Input blower settings Power consumption Pitot tubes Orifice plates	Determine approach velocity (m s⁻¹), empty bed detention time(s), true detention time(s)
Contaminants and products		
Input, output (ppmV or g m⁻³)	GC, H₂S meter, IR, detectors (FID, PID, DELCD, etc.) Olfactometry	Calculate loading rate, elimination capacity; estimate biological heating contribution to evaporation
Products: concentration in effluent, medium, or leachate (ppmV or g m⁻³)	GC, HPLC, GC/MS	Look for toxic by-products, check mass balance
Carbon dioxide input, output, profile (ppmV or g m⁻³)	GC analysis, ND-IR	Determine respiration rates, close mass balance
Concentration profiles (ppmV or g m⁻³)	GC, H₂S meter, detectors, multiple determinations	Determine treatment kinetics (FID, PID, DELCD, etc.)
Variation of input (ppmV or g m⁻³ over time)	GC, H₂S meter, detectors (FID, PID, DELCD, etc.), multiple or continuous determinations	Look for short-term overloads
Pulse test for retardation factor (peak velocity, cm s⁻¹, peak size, g m⁻³)	GC, H₂S meter, detectors (FID, PID, DELCD, etc.), multiple determinations	Calculate mass adsorbed from peak velocity; determine degradation rate from decline in peak size
Medium characteristics		
Particle size distribution	Sieving	Determine compaction tendency
Moisture characteristic curve (plot potential vs. water content; %)	Lab experimentation with tensiometer	Determine range of acceptable moisture contents
Field capacity (% H₂O)	Lab experiment; add water until flow occurs and measure weight gain	Determine maximum moisture content
pH, pH profile	Moisten, pH meter, or pH paper	Check for acidification, acid inhibition, or acid damage to medium

Table 8.1 (continued)

Parameter	Method	Purpose
Buffer capacity (mol L^{-1} or mol g^{-1})	Lab titration	Determine biofilter resistance to acidification
Medium organic and ash content	Dry at 90°C; ignite at 550°C	Determine changes in organic content and degree of degradation
Moisture content		
Relative humidity input and output (%)	RH sensor, wet and dry thermometer	Determine tendency of biofilter to dry
Medium moisture content (% H$_2$O)	Lab gravimetric determination, tensiometer and characteristic curve, load cells, moisture probes	Determination of medium moisture content
Amount of irrigation water added, (L h^{-1} m^{-2})	Flow rate and duration	Watch for overwatering, check water balance
Temperature		
Input/output temperature, temperature profile (°C)	Thermometers, temperature probes	Look for temperature inhibition; determine heat balance; understanding of moisture control
Temperature variation with time (°C over time)	Thermometers, temperature probes, multiple determinations	Look for temperature shock
Air permeability and permeability homogeneity		
Pressure drop (head loss) across bed (cm H$_2$O or psi), decline in flow rate for fixed power or speed blower system	Input and output pressure gauges, difference manometer, flow meters	Determine degree of compaction or clogging
Smoke test	Insert smoke bomb in biofilter inlet	Check for flow nonhomogeneity
Local velocity determination	Ducted anemometer	Check for flow nonhomogeneity
Methane pulse, for flow pattern	Add methane pulse, GC at outlet	Determine true air flow retention time; check for flow non-homogeneity
Biological activity		
Treatment success	Analysis of input and output contaminant concentrations	Check microbial ecosystem health

Table 8.1 (continued)

Parameter	Method	Purpose
Microorganism counts (number g^{-1} medium)	Plate counts, fluorescence microscopy	Check microbial ecosystem health
Qualitative assessment of microbial ecosystem (approximate number g^{-1} medium)	Microscopic inspection	Determine biofilm structure, determine some components
Pollutant degradation rate (mg g^{-1} medium h^{-1})	Laboratory test with medium samples in closed containers	Determine specific pollutant degradation rate
Respiration (mg CO_2 g^{-1} medium h^{-1})	Lab test with medium samples in biometers	Check microbial ecosystem health
Biomass, (g g^{-1} medium)	COD or BOD (for inorganic media only)	Check microbial ecosystem health
Fatty acid analysis (mg g^{-1} medium)	Extraction and GC analysis	Determine microbial ecosystem characteristics
Leachate characteristics		
Flow rate	Measure volumes accumulating in leachate collection	Determine need for more or less irrigation, detect excess condensation
pH	pH meter, pH paper	Look for biofilter acidification
Contaminant concentration (mol L^{-1})	GC, GC/MS	Determine loss of contaminant to leachate
Particulates (g L^{-1} in size class)	Sieving	Determine decomposition of medium
Salts	Optical density, polarography, conductivity	Look for production of chlorides, sulfates or carbonates
Biological or chemical oxygen demand	BOD or COD standard tests	Determine means to dispose of leachate, biodegradability of leachate
Total carbon	TOC analyzer	Determine mass of carbon in leachate for carbon balance
Biofilter medium		
Heavy metals (ppm)	Atomic absorption spectroscopy	Test for toxics prior to medium disposal
Toxic organics	GC/MS, GC, HPLC	Test for toxics prior to medium disposal

Table 8.1   (continued)

Parameter	Method	Purpose
Pathogens (CFU g^{-1})	Plate counts on selective media	Test for presence of pathogens prior to medium disposal
Visual inspection		
Surface		Look for overdrying, overwetting, medium decomposition, apparent fungus or biofilm
View ports with cleaning procedure		Look for overdrying, overwetting, medium decomposition
Humidifier		
Slime accumulation	Visual inspection, BOD in purge water, pressure drop	Avoid clogging and head loss due to slime buildup
Salt accumulation	Conductivity in purge water	Avoid precipitation of minerals
pH	pH meter or pH paper	Avoid corrosive conditions
Input and output relative humidity	RH meter	Assess overall humidifier performance
Input and output temperatures for water and air	Thermometers, temperature probes	Assess effects of temperature on humidifier performance

Note: GC = gas chromatography, HPLC = high performance liquid chromatography, GC/MS = gas chromatography/mass spectrometry, ND-IR = non-dispersive infrared spectrometry, FID = flame ionization detector, PID = photoionization detector, DELCD = direct electrolytic conductivity detector.

Source: Adapted from by Devinny, J.S., *Hazardous, Toxic and Radioactive Waste Practice Periodical*, 2(2), 78–85, 1998. With permission.

the flow resistance in the biofilter and reduce air flow rates, even if blower operation does not change.

Operators should know the air flow rates in order to understand the effects of pollutant load. Additional pollutant load may increase the tendency for biomass accumulation or pH decline. Increasing air flow will cause more evaporation, and water additions should be reduced during periods of low air flow to avoid saturating the bed. Sufficient air flow should be maintained during down times to prevent development of anaerobic conditions. In the case of lightly loaded systems, overnight shut-downs may be acceptable because the amount of contaminant present will not rapidly deplete the oxygen present in the pores.

For many biofilters, instruments which are read manually are sufficient. The necessary frequency of measurement depends on the degree of flow variability. If flows are stable, data can be recorded daily, or even at longer intervals. For highly variable flows, it will be desirable to collect data more often using automatic systems.

Field measurement of air flow rates is a well-established technology, and many kinds of flow meters are available. They may use pitot tubes, turbines, orifice plates, or heat loss from an exposed element to determine air velocity. The choice for a particular application depends on pipe size and air flow rate. The primary condition which is special to biofilters is the necessity for resistance to corrosion. The input air may contain sulfide, which will corrode many ferrous alloys, and organic acids generated in the biofilter are sometimes present in the output.

8.4 Contaminants

8.4.1 Input and output concentrations

The amount of contaminant being removed in a biofilter is the primary measure of its effectiveness. The operator must monitor this to determine whether the system is working and is often required to do so in order to meet regulatory requirements. The operator should know how much contaminant is entering the biofilter in order to anticipate operational needs. An example of the importance of this measure is one case in which the biofilter owners asked for help in biofilter monitoring, and no contaminants were found to be entering. The upstream pollutant control process was working so well that there were no releases. The owner is now planning to shut down the system.

Experimental work has frequently included measurement of concentration as a function of depth within the biofilter. Choosing the appropriate model is vital if experimental results are to be applied to the design of full-scale systems. The characteristics of the depth profile may also indicate whether treatment is limited by biological activity or diffusion in the biofilm. There are additional difficulties involved with sampling within the biofilter, however. Flow heterogeneity within the medium may make concentrations different even at the same depth within the medium. Tubes or other devices built into the medium may cause systematic disruption of the flow in their immediate vicinity, magnifying the effect, and biomass may clog the ports. If the syringe needle inserted to take the gas sample inadvertently touches the biomass, a bit of it may be inserted into the gas chromatograph, giving erroneously high results.

In laboratory systems, biofilters are often divided into sections separated by air-filled gaps. Within the gaps the gases can remix, so that the effects of any particular heterogeneity are limited to one section, and the measurement on gas taken from the gap represents the average effects of the preceding biofilter section.

8.4.2　Contaminant concentration variability

Biofilter performance is affected both by average input concentrations and by their variability. Just as air flow rates rise and fall with variations in source activity, concentrations will change. The processes that generate the contaminant may be periodically shut down, and releases from batch processes may be sudden. Microorganisms work best when concentrations are constant. Periods without contaminant will gradually cause the ecosystem to lose its ability to degrade the compound. Spikes of high concentration may be toxic. Biofilters generally respond well when periods of low contaminant concentration last for only a day or two. The effects of longer outages can be mitigated by providing the biofilter with a low air flow and even an artificial contaminant feed (Kinney et al., 1996a). If monitoring indicates that variability is a problem, input concentrations may be stabilized by process venting modifications, or by installation of adsorbent pretreatment.

Input concentrations may vary over periods as short as hours or minutes, and such high frequency changes will only be fully described by automatic concentration monitoring. For biofilters treating mixed discharges from variable processes, complete characterization may never be possible.

8.4.3　Pulse testing

Monitoring of variation in input and output concentrations may provide useful information about biofilter operation. Because biofilters have a substantial adsorptive capacity, a pulse of contaminant which enters a biofilter will pass through it at a velocity less than the velocity of the air, just as a peak moves through a gas chromatograph column (Hodge and Devinny, 1995). If the velocity of the peak is measured and compared to the air velocity, the adsorptive capacity of the bed can be calculated. Such an experiment is most likely useful in experimental evaluation of biofilter media at the bench scale, but it might also be useful for evaluation of medium deterioration. The test requires the ability to measure concentrations at the input and output at high frequency, so the shape of the peak can be defined.

Because variations in input concentration are propagated through the biofilter with some time delay, caution may be necessary in interpreting monitoring results for removal efficiency. If input and output concentrations are measured simultaneously, and the input is variable, it is possible that one measurement will catch a peak concentration while another is made between passing peaks. Removal efficiencies can be substantially exaggerated or underestimated. Ideally, the output sampling should be done after the input sampling, and the time between samples should be equal to the contaminant detention time. Alternatively, care should be taken to measure removals during periods when the input concentration is not changing or to average them over long time periods.

8.4.4   Product concentrations

Biofilters which are overloaded may generate products which are carried out of the biofilter before degradation to mineral products is complete (Devinny and Hodge, 1995). These emissions may be illegal or undesirable and indicate biofilter problems which should be corrected. At least initially, detection of such products requires analysis by gas chromatography/mass spectrometry (GC/MS), because their identity will not be known. If the problem is continuing, an initial identification of the peaks may allow most subsequent determinations to be made by gas chromatography.

In the most elaborate analyses, input and output concentrations can be used to develop a mass balance for the biofilter. The amount of carbon entering can be calculated from input concentrations and flows. Output can be calculated from the flows and concentrations of untreated contaminant, by-products, and carbon dioxide, in the air and in the leachate. If less carbon is being released than is coming in, biomass must be accumulating in the biofilter, and clogging may be imminent (Medina et al., 1995a; Auria et al., 1996). Kinney et al. (1996) used carbon balance calculations to show that frequent alternation of the direction of flow in a biofilter could suppress biomass accumulation.

8.4.5   Instrumentation and techniques

The sophistication of contaminant monitoring systems varies widely. When asked about output monitoring for treatment plant odors, one operator said that if there are no complaints from the neighbors, the biofilter is working. For odors, which come in great variety and at very low concentrations, the human nose is often the best instrument. Quantitative measurements can be made by odor panels. These groups of individuals sniff samples of air to determine whether an odor is present. They are provided with samples which have been diluted to various degrees, so that in some the odor is no longer perceptible. The concentrations of odor-causing contaminants are often measured in odor units per cubic meter (OU m^{-3}), equal to the dilution factor required to reduce the concentration of the contaminant to its odor threshold.

Gas chromatography and GC/MS are the most common instrumental methods for measuring contaminant concentrations, and there are many specific protocols for column type, flow rate, and temperature program which have been developed for other applications in air pollution control. Measurements for biofilters are made in the same way. Hydrogen sulfide meters and other special purpose instruments are available for specific compounds.

The decisions to be made for biofilter monitoring concern the frequency and sophistication of the analysis. Grab samples may be taken using Tedlar® sampling bags or evacuated stainless steel cylinders and then transported to

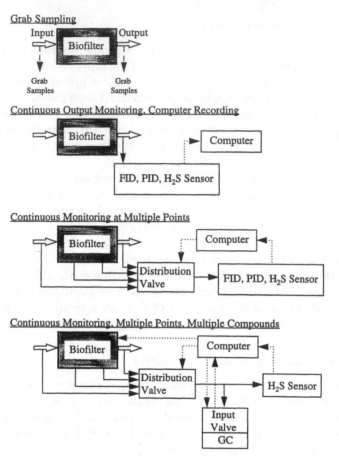

Figure 8.2 Example of sampling, analysis, and control systems for biofilters. Solid arrows represent air flows; dotted arrows represent the flow of data or control information.

the laboratory. The laboratory can then readily do GC, GC/MS, or a variety of other analyses. However, the sampling and analysis costs per sample are likely to be high, and the approach is workable only when the sampling frequency can be low. Automatic on-site instruments generally become economical when the sampling frequency is greater than once per week. Special-purpose meters can be operated continuously. Air can also be passed continuously through a flame ionization detector, or some of the other detectors commonly used for gas chromatographs. However, because it is usually desirable to sample from at least the input and output and possibly at other points in the bed or from multiple biofilters, automatic equipment for pumping the samples to the instrument and switching to analyze flows from various sources is necessary (Figure 8.2).

A flame ionization detector alone, however, cannot distinguish among the compounds it detects. It will be adequate for measurements in a system treating a single compound but will not provide data on the various components of a mixed contaminant stream. It is often used to measure overall efficiency, characterized as a reduction in volatile organic compounds; however, operators should be aware of bias that arises because the detector is not equally sensitive to all organic compounds. Standefer (1998) has noted that the flame ionization detector is less sensitive to oxygenated compounds that are often most efficiently treated. For a biofilter treating a mix of oxygenated compounds and hydrocarbons, actual removal efficiencies will be higher than indicated by the decline in the signal from a flame ionization detector.

To measure removal efficiencies for individual compounds (and by summation, the true overall efficiency), a gas chromatograph can be installed to operate automatically. Chromatography will separate the contaminants and allow their individual concentrations to be determined but requires another increase in monitoring instrument sophistication. A gas chromatograph cannot operate continuously. It passes a discrete air sample through the adsorptive column to separate the components, often while the column is being heated according to a precise program, and the signal from the detector at the output of the column creates the chromatogram. When the data are collected and the instrument has been returned to its starting condition, another sample can be accepted. Thus the GC must have the ability to draw in precise samples at programmed intervals, in addition to being connected to a pump and automatic valve system which provide samples from various points in the biofilter at specific times.

A system with flow switching and programmed sampling could also be used to operate a system using a gas chromatograph/mass spectrometer at the site, providing direct identification of compounds. If the pollutants have distinct mass spectra, it is conceivable to install a mass spectrometer on-line and perform continuous monitoring of single masses. Such systems have been used for monitoring bioreactors in biosynthetic systems. They allow, for example, simultaneous monitoring of oxygen, water, and selected compounds. However, both GC/MS and MS are expensive and require frequent maintenance and extensive operator training, and there are no reports of installation of such a system in the field.

To record the data from many instruments, to manage the sampling system, and often to control some aspects of biofilter operation, a programmable logic computer (PLC) system may be employed. Despite the complexity of the task, systems exist for relatively easy application. Laboratory management software installed in a personal computer can record data, open and close appropriate valves, turn pumps on and off, and shut the biofilter down if instruments indicate trouble.

Many contaminant monitoring systems require installation of tubing which leads from the sampling point to the instruments, in order to draw gases into the detector. These are a common source of problems. Water frequently condenses, and droplets within the sampling lines will adsorb or desorb contaminants as concentrations change, confusing the results. Biological growth may occur, and it or particles of medium may clog the lines. Operators should be particularly suspicious of unusual determinations of zero concentration, which may indicate blocked sampling lines rather than good biofilter performance. Such problems may be greatly reduced by periodically flushing the sampling lines with ambient air. Heated sampling lines are available and effective, but are expensive.

8.5 Medium characteristics

8.5.1 Particle size distribution

While it has not been widely reported, it seems reasonable that biofilter media should be monitored for changes in mean particle size and particle size distribution. As compost biofilter media age, they continue to biodegrade. One effect is that large particles are broken up. Occasional measurements of the particle size distribution would provide warning of the impending need to replace the medium. These can be readily measured by screening a sample of the medium in standard soil sieves.

8.5.2 Moisture characteristic curve and field capacity

The moisture-holding characteristics of biofilter media can be described by a "characteristic curve" relating the chemical potential of the water to water content (Chapter 4). In practice, determining the entire characteristic curve is difficult, and the field capacity of the medium is often taken as a measure of its water-holding ability. This may be misleading in the case of media with substantial porosity in the size range too small for microorganisms. Much of the water they hold is not available to microorganisms.

The characteristic curve is determined by measuring the soil water potential at various water contents, typically using a tensiometer. Field capacity can be determined by saturating the medium, allowing it to drain, and then determining the weight loss on drying at 100°C.

8.5.3 pH and alkalinity

Different biofilters may have different ideal pH values, depending on the contaminant being treated and the characteristics of the microbial ecosystem; however, changes in pH generally stress the microorganisms. Because acidification is most intense near the inlet of the biofilter, a sample from this region should be checked first. Occasional profiles with depth may reveal information on which parts of the biofilter are most active. Samples of medium from several depths in the biofilter should be tested for pH at regular intervals, with the length of the interval depending on the compounds present in the air and the efficiency of the biofilter.

Determination of pH is most precisely done by mixing a sample of the medium with distilled water and measuring the pH of the water with a pH meter. For routine checks, however, it is sufficient to press a sample of the medium against pH paper and note the color change.

Biofilters that are treating air flows with significant concentrations of acidifying gases will require a means for removing or neutralizing the acids. Base may be added at intervals with the irrigation water. For biofilters subject to low levels of acidification, it may be sufficient to have basic materials such as limestone or crushed oyster shell included in the medium. As these media age, the basic materials are gradually exhausted. Periodic titration of samples with acid will determine their remaining alkalinity and warn operators of the incipient need for medium replacement. Changes are usually gradual. As experience is gained with a specific situation, the operators can estimate the necessary sampling frequency from the measured rate of change and the total pH change that is considered acceptable.

Leachate pH should also be measured. It will reflect conditions near the bottom of the bed. For upflow biofilters, this indicates what is happening at the inlet. For biofilters with lower soaker hoses, it may provide information to guide selective leaching of inlet zones.

8.6 Water content

8.6.1 Input and output relative humidity

Proper control of water content may be the most critical aspect of biofilter operation. Unfortunately, relative humidity measurement in the range important for biofilters is surprisingly poorly developed. Biofilters must operate in the water content range between field capacity and the minimum necessary for high biological activity. The relative humidities of air at equilibrium with media in this range, however, are only slightly below 100%. Small changes in relative humidity ultimately produce very large changes in medium water content, so relative humidity must be measured quite accurately. There are many instruments available for measuring relative humidity. Historically, comparing wet and dry bulb temperatures to psychometric charts or using hair hygrometers were the only means. These techniques are still widely used, even though resistive or capacitance meters have largely surpassed these in terms of response time and accuracy of measurement. Even so, most techniques are still not very sensitive in the range very near 100%, which is of interest for monitoring biofilters. While no one has yet reported such a procedure, relative humidity (RH) near 100% might be measured by a conventional RH meter after heating the air sample by 10 or 20°C. The relative humidity of the initial sample can then be calculated using phychometric tables and the measured absolute humidity. This procedure might prevent the problems usually associated with the measurement of nearly saturated air streams while allowing the use of less expensive humidity meters.

8.6.2 Medium moisture content

Samples of medium can be removed from the biofilter for determination of the loss of weight on drying to 100°C. This method is accurate and provides the information desired within a day or two, but it requires manual collection of medium samples. Biofilters tend to dry most rapidly near the inlet, so profiles of water content as a function of depth are desirable. Drying can also vary from place to place over the area of the biofilter. A complete determination requires that samples be taken at several depths and locations throughout the biofilter (van Lith et al., 1996). Care must be taken to refill the resultant holes, so that air flow in the biofilter is not short-circuited. Overall, the process is effective but labor intensive.

Biofilters contained in tanks are sometimes mounted on load cells to provide continuous measurement of the weight of the biofilter. Because changes in water content are the primary cause of weight changes, the cells provide a means for accurate continuous water content monitoring. Some drift may occur as biomass is accumulated or medium is degraded, so occasional recalibration is warranted. High-quality load cells are necessary to provide accurate determination of the medium moisture content. The changes in water content that are of interest to the operator are relatively small in comparison to the total weight of the water, medium, and reactor structure. For example, if one cubic meter of medium weighing 400 kg m^{-3} and supported by 100 kg of structure includes 60% moisture, a change in weight of only 20 kg represents a decrease in moisture content from 60 to 58%. To measure this change, load cells must be capable of accurately measuring a 4% change in the total weight.

Water content in soils can be determined by time domain reflectometry, and such sensors would likely be effective for biofilters; however, they are relatively expensive ($5000 to $10,000) and have not yet seen use in the field. Tensiometers, which measure soil water suction directly, have also found wide application in soils. Some attempts have been made to use them with biofilters, but the large size of the biofilter particles means that contact with the tensiometer sensing surface is poor, so the response time is long and results have been suspect. They share the disadvantage with many sensors that they measure only the water content in the immediate environment of the sensor and will not detect drying occurring in any other part of the biofilter.

8.6.3 Irrigation water

The operator controls the amount of water added to the biofilter by the irrigation system. In general, the water content should be measured frequently, and the rate of water addition should be adjusted until it maintains water content at the chosen value. The rate of addition is commonly monitored indirectly, as the amount of time that watering is done. The time is adjusted as experience shows it to be too much or too little. This may be as simple as timing the worker who is spraying the surface of a simple biofilter with a hose. More commonly, installed sprinkler systems are operated by a timer and can be adjusted to operate for a chosen number of minutes per day. If the bed begins to dry or higher air flows are expected, the number of minutes can be increased. However, it should be kept in mind that the amount of water delivered by sprinklers can vary. Sprinklers are frequently partially clogged, reducing flows. Readjustment of the sprinklers will change the flow. It may be valuable to have a flow meter on the water delivery system to detect such changes, and sprinkler nozzles should be checked at least twice a month.

As was described for startup procedures, it is important to maintain the uniformity of water distribution. Sprinklers may clog. Individual sprinklers should be inspected, and occasional measurements of sprinkling uniformity should be made.

8.7 Temperature

Temperature measurements in biofilters are common and have at least two uses. Monitoring the incoming air temperature will warn the operators if the source process is producing air hot enough to kill the organisms or cold enough to inhibit their activity. Alteration of the source process or mixing of air streams may allow some control of input air temperature. Continuous temperature records may allow operators to identify events which have disrupted biofilter performance.

Temperature is also important for control of water content. If the incoming air is warm and dry, it will dry the biofilter. If it is warm and humid, and the biofilter is out in the cold, there may be continual condensation and saturation of the medium. Measurement of the difference between wet and dry bulb temperatures, of course, is one means of determining relative humidity.

Temperature change within the biofilter can be used as a measure of biological activity in the biofilter (Medina et al., 1995a; Auria et al., 1996). Metabolic heat will warm the air noticeably and can contribute substantially to bed drying (van Lith and Leson, 1996). Measurement of temperatures in the incoming and outgoing air and at intervals within the bed can readily be done with thermocouple sensors. These can be wired to meters for manual record keeping or to a computer for automatic recording.

8.8 Medium permeability

8.8.1 Average permeability and head loss

The input to output pressure difference (pressure drop or head loss) required to drive air through the biofilter is an important operating parameter. The pressure drop in a biofilter rises as permeability is reduced, so it can be used as an indirect measure of bed permeability. For most kinds of blowers, increases in pressure drop will cause reductions in air flow rates. Any measurement of head loss must be accompanied by a flow rate measurement if it is to be interpreted in terms of the quality of the biofilter medium. Records should be kept so that comparisons under similar flow conditions can be made. Sharp increases in head loss and declines in flow rate may signal the need to till or replace the bed material.

Pressure gauges should be installed immediately above and below the medium at the input and output so that they accurately measure pressure drop across the bed. For many applications, a water-filled u-tube manometer connected to inlet and outlet ports may be quite adequate. Many instruments are available which can be read manually or which will report to automatic recorders. In the case of open bed biofilters, the comparison will be made between the inlet pressure and atmospheric pressure. Care should be taken to ensure that the pressure determination is not influenced by air flow rates at the point of measurement. Pressure sampling lines should be checked weekly for buildup of biomass, clogging by particles, or condensation.

8.8.2   Permeability homogeneity

If the resistance to flow in the biofilter medium varies strongly from place to place, the air flow will short-circuit, drastically reducing actual contact time and treatment efficiency. Measurements of the overall pressure drop may not show such effects. If the air is flowing through fissures and channels, pressure drop may be low even though much of the bed has substantially reduced permeability and treatment is poor.

For open-bed biofilters, a smoke bomb placed in the air input line can provide a quick qualitative test of flow homogeneity. Smoke in the air flow allows the operator to see where air is flowing preferentially through the bed. It is often particularly noticeable that air flows preferentially along the walls of the biofilter. This observation has led to the installation of baffles on the walls of some systems (Chapter 7).

In some experimental systems, investigators have checked flow uniformity by introducing a pulse of trace gas into the air flow (Deshusses, 1994; Sabo, 1991). An inert, easily detected tracer that is not significantly sorbed on the packing should be chosen, such as methane or propane. Continuous monitoring of the outflow after a pulse in the inlet shows the shape of the concentration peak as it leaves the biofilter. Peak broadening is primarily the result of two phenomena: axial diffusion and heterogeneous flow. While axial dispersion will broaden the peak in a symmetrical manner and can be quantified using dispersion theory and the Bodenstein dimensionless number, heterogeneous flow will produce asymmetric deviation. For example, significant tracer detection in the outlet before the true gas detection time is a clear sign of short-circuiting. In a similar manner, a long peak tailing after the true bed detention time is an indication of dead zones within the biofilter bed. Both short-circuiting and dead zones may occur at the same time. In any case, if such observations are linked to poor reactor performance they may indicate a need for reworking or replacing the biofilter medium.

Direct measurements of flow homogeneity are possible. Output air flows can be measured locally by using a ducted anemometer at various locations on the top of the bed so that flow uniformity can be assessed.

8.9   Biological activity

8.9.1   Treatment success

Microbial activity is the basis of biofilter function. Maintaining it is a principle objective of the biofilter operator. If other operating conditions are good, the removal efficiency of the biofilter is itself an excellent measure of microbial activity, and often no independent measures are needed.

Biofilter effectiveness may decline for other reasons, however, and often it is useful to have a direct means of monitoring the health of the microbial ecosystem. An operator, for example, may be unsure whether poor treatment is occurring because the microorganisms are inactive or because short-circuiting is occurring. The operator might also wish to assess microbial activity in different parts of the biofilter, to determine whether local differences in medium quality are important.

8.9.2   Cell counts

The number of microorganisms present may provide useful clues. Plate counts of heterotrophic organisms are readily made on small samples of medium. Heterotrophic counts use general media and provide an overall measure of microbial activity. Counts on special media, containing only the contaminant of interest, will enumerate organisms capable of growing with the contaminant as their sole carbon source. However, plate counts are labor intensive, and the variance in the determinations is relatively high. The analysis takes a week to 10 days because the organisms must grow enough to create visible colonies for counting.

The results must also be viewed with caution because of their lack of completeness. Microbiologists are now using sophisticated genetic methods to show that there are many more species of microorganism in soils than can be grown in petri dishes. Various estimates are that only 10%, 1%, or even fewer of the species present can be grown in a laboratory (Amann et al., 1995). It is likely that there are many species that can degrade the contaminant of interest in the rich environment of the biofilter medium but cannot do so in a petri dish with only the contaminant and mineral nutrients present. Webster et al. (1997) noted that plate counts of microorganisms made in a biofilter were lower than those estimated from phospholipid fatty acid analysis and that the discrepancy increased as the pH in the medium declined, making it more and more different from the conditions in the petri dish. Even so, plate counts are useful as relative measures. If the counts decline with time, it probably indicates trouble. If they are lower in particular environments, such as near the inlet or in dry spots, they may suggest why the trouble is occurring. Reasonable values are 10^8 to 10^9 CFU g^{-1} for total heterotrophs and 10^7 to 10^9 CFU g^{-1} for organisms that can grow on the pollutant as a sole source of carbon.

Microscopic inspection of biofilter media is less often used because it is difficult to prepare specimens. Most of the cells cling tightly to the medium particles. These block the light so that ordinary microscopy, which depends on light passing through the specimen, sees only blackness. Methods using fluorescent stains have been used experimentally. These stains can make cells stand out against a black background and may distinguish between living cells and small particles of medium. In experimental programs, Hugler et al. (1996) examined biofilm cross-sections to gain insight into the operation of a biotrickling filter.

8.9.3   *Respiration*

The activity in a specimen of biofilter medium can be measured using respirometry (Acuna et al., 1996; Hodge and Devinny, 1997). A sample of the material may be placed in a flask which is connected to a reservoir of KOH solution. Any carbon dioxide released dissolves in the solution, and the amount can be determined by back titration of the solution. The rate of carbon dioxide release per gram of medium is a measure of overall activity. More elaborate automatic systems are available which measure respiration as CO_2 release or O_2 consumption in tests lasting from 1 to 4 days. The primary difficulty with respirometry is that the sample is isolated from the biofilter and from the flow of contaminant. Results will depend on the details of the experimental protocol, including sample water content, nutrient supply, pollutant supply, and others, and care must be taken to ensure that the test reproduces the activity in the biofilter.

8.9.4   *Biomass*

The amount of biomass present may reflect biological activity. If biomass is accumulating rapidly, it means the organisms are healthy and treatment is rapid, but it also means that the biofilter may soon be clogged. Biomass cannot be measured on organic matter such as compost, because it is not possible to distinguish between the biomass of the active organisms and the biomass of the compost. On inorganic media, however, biomass can be measured by drying and weighing a sample, then heating it to 550°C to determine total volatile solids. Because biomass is itself biodegradable, another approach is to kill the biomass on a sample, suspend it in water, and measure its concentration as biological oxygen demand.

Other measures may at least partially distinguish between living and dead biomass. The protein (Cox et al., 1997) or adenosine triphosphate content of the material may be determined, and newer methods may measure the DNA content.

8.9.5 Fatty acid analysis

Fatty acid analysis has been used in some biofilter studies. Many varieties of phospholipid fatty acids are found in the membranes of microorganisms, and some generalizations can be made about the kinds associated with particular groups of microorganisms or the conditions under which they are growing. Certain classes of compounds, for example, are associated with gram-negative bacteria, rapidly growing bacteria, or microorganisms under environmental stress. The fatty acids can be extracted from a bulk sample and analyzed by gas chromatography, allowing general characterization of the microbial community present. Webster et al. (1997) showed that the microbial community is still changing after hundreds of days of biofilter operation, and that microorganisms were stressed in biofilters where the pH declined. de Castro et al. (1996) found that biofilters seeded with different inocula supported somewhat different microbial ecosystems even after long periods of operation and even though both were treating the contaminants effectively. Phospholipid fatty acid analysis, however, is labor intensive and expensive and yields relatively general results of more interest to the researcher than the operator. Until more specific relationships are understood, it will likely remain a research tool.

8.9.6 Visual inspection of biofilters

Perhaps the most common type of biofilter monitoring is a visual inspection. It is certainly inexpensive and can have substantial value if the operator is experienced. The appearance and feel of the medium may immediately tell the experienced operator whether it is too wet or too dry and whether watering is uniform. A sample dug from the biofilter with a trowel may show obvious accumulation of biomass or other signs of clogging. If many small particles are present, and the larger pieces of compost are soft and tightly compacted, the material may be due for replacement. While it has not been reported, it would be simple to install "measuring sticks" — vertical rulers in the compost — and observe the depth of the medium. As compaction occurs, the bed will become shallower. Short-circuiting and overwatering may create anaerobic zones in the biofilter, and these could be detected by their odor if the operator examines the surface of the bed closely.

8.10 Leachate characteristics

Biofilters are intended to operate with a stationary water phase. Drainage may occur inadvertently, however, when there is rain on an open biofilter, when condensation from the input air is unusually heavy, or when irrigation is overdone. Biofilters may occasionally be overwatered intentionally to wash out acids or salts. The leachate may be analyzed by any of the great number of methods that have been developed for water and wastewater (refer to *Standard Methods*, American Public Health Association, 1995).

pH is commonly measured in biofilter leachate, often simply because it is easier to measure leachate pH than medium pH. It should be noted, however, that the leachate pH reflects conditions near the bottom of the biofilter. In a downflow biofilter, the pH may be low at the top and neutral at the bottom. Water flowing downwards will be neutralized by passage through the lower layers of medium.

Nutrients are commonly added to biofilters and will increase the ionic strength of the water phase. Several recent publications have indicated that too much nutrient can decrease microbiological activity in porous media, presumably because the ionic strength in a modest volume of water is changed significantly (Brook et al., 1997). Measurements of leachate salt content might tell operators when flushing the medium is advisable.

Leachate can also be checked for concentrations of the contaminant and by-products. This may be necessary for proper leachate disposal. It is desirable that the leachate not require disposal as a hazardous waste. Disposal to wastewater collection systems requires that the leachate meet certain quality requirements, and high contaminant or by-product concentrations may prevent this. By-product concentrations may also be diagnostic of biofilter operation.

Microbial degradation of compost produces humic acids. It is interesting to speculate that degradation of organic biofilter media might be monitored by measuring humic acid concentrations in the leachate.

8.11 Humidifiers

Many biofilters systems include humidifiers designed to provide air to the reactor at the highest possible humidity. These are commonly spray chambers or packed towers which provide a large water surface in order to promote evaporation. Monitoring is necessary to assure that these remain in good working order.

The humidifier brings large amounts of water in contact with the contaminated air. Although nutrient content and support material are not optimized for microorganisms as they are in the bioreactor, biological growth may occur. Slime buildup can clog fittings and, in the case of packed towers, may build up to the point where the pressure drop across the humidifier increases. Monitoring of the pressure drop and visual inspection will detect the problem. The biomass can be removed through application of bactericides such as chlorine. In spray chambers, the performance of the spray nozzles should be checked frequently to assure that clogging is not causing a uneven spray distribution.

Most humidifiers recirculate the water in order to avoid excessive water use. Some purge is necessary, however, in order to avoid salt or acid buildup as water evaporates and is replenished. Humidifiers will also collect particulates from the air. Total suspended solids and conductivity can be measured to determine whether the purge flow rate is adequate. Measurement of BOD will provide warning of biological activity. The pH of the purge should also be monitored to check for biological activity and to warn of corrosive conditions.

The overall performance of the humidifier is determined directly by measuring the relative humidity of the incoming and outgoing air. If the effluent humidity is not as high as desired, input and output water temperatures may suggest why. If both the air and water are cold, evaporation rates will be low, and some form of temperature control may be necessary.

8.12 Statistical analysis

Input contaminant concentration, air flows, temperatures, and many other factors that control biofilter operation are variable. This can make evaluation of even a simple parameter, such as removal efficiency, difficult and subject to error. Typically one instrument is used to measure input and output concentrations so the determinations cannot be made simultaneously. Trends in biofilter performance, such as gradually increasing pressure drop, may be masked by variability. While it is not possible to include a survey of statistical techniques here, it is important to note that biofilter assessment requires appropriate methods of statistical inference, just as other engineering studies do.

8.13 Conclusions

Biofilters rely on a fundamentally simple concept, but treatment eventually results from the activity of a microbial ecosystem. Like all ecosystems, it is governed by complex physical, chemical, and biological factors. Good biofilter operation requires understanding of the ecosystem and monitoring of the controlling factors. It is possible to compile a long list of parameters that could be valuable. Frequent determination of all of the values, however, would be impossibly expensive. The challenge for biofilter monitoring is to choose a shorter list and a sampling frequency for each that are appropriate to the application, provide sufficient guidance for good management, and are economical.

Monitoring is particularly important during the startup period, when the microbial ecosystem makes the transition from a low-density inoculum to a thick, well-acclimated biofilm. Understanding of the basic principles and collection of the necessary data can help ensure rapid and successful establishment of treatment capacity.

chapter nine

Application of biofilters

9.1 Introduction

In this chapter, applications of biofilters are presented. Every biofilter in the world is unique by its construction or by the type of air it is treating; however, the 13 applications discussed in detail in this chapter were chosen because they represent typical examples of field applications of biofilters. These case studies are reported to the best of our knowledge, based on published information and personal communications. Their description in this book does not constitute an endorsement of the biofilter design or of the biofilter vendor. Note that the treatment cost calculation was often through personal communication and that the method used to calculate the cost may be different from case to case. Hence, treatment costs may not be directly comparable.

There is much to learn about past successes and failures. In general, those who fail to learn from history's errors are condemned to repeat them. Certainly, there is some truth in Oscar Wilde's quotation "Experience is the name everyone gives to their mistakes."

9.2 The ARA-Rhein biofilter: wastewater treatment odors and VOCs

Various streams of waste air loaded with odors and low concentrations of VOCs were treated in a large biofilter at the wastewater treatment plant of a fine chemical production facility (Table 9.1; Figure 9.1).

9.2.1 Design and operation comments

The original pre-humidification system, consisting of three packed-bed humidifiers in parallel, rapidly clogged because of biomass growth and because

Table 9.1   Characteristics of the ARA-Rhein Biofilter

Owner and location	Novartis (formerly Sandoz and Ciba-Geigy), ARA-Rhein; Basle, Switzerland
Builder	Monsanto Enviro-Chem (MEC)/ClairTech BV; Woudenberg, The Netherlands
Type of air stream	Exhaust air from wastewater treatment plant
Year of installation	1990
Medium type and volume of medium	BIOTON® (ClairTech BV, Woudenberg, The Netherlands) (50:50 compost and polystyrene spheres), 2100 m³; extra limestone added to buffer pH drop due to chlorinated compound degradation
Number of layers and height of medium	Three parallel filters of 3 layers of 1 m height each Downward air flow
Biofilter construction type	Closed biofilter made of a rectangular 304 stainless steel (some 316) box of 35 m by 20 m, 6 m in height, thermally insulated
Humidification	Pre-humidification and water spray over the filter bed (18 sections controlled independently over each layer of medium)
Air flow rate	60,000–75,000 m³ h⁻¹ (35,000–45,000 cfm) split in three lines
Empty bed residence time	1.5–2 minutes
Pressure drop	First layer 20 cm of water gauge per meter of medium, second and third 10 cm wg per meter
Average bed temperature	20–35°C
Pollutants treated	Initially: Chlorinated hydrocarbons: 25–50 mg m⁻³ Aliphatic hydrocarbons: 50–100 mg m⁻³ Aromatic hydrocarbons: 200–250 mg m⁻³ Unknown hydrocarbons: 50–100 mg m⁻³ Identified compounds were toluene, xylene, methanol, isopropanol, chloroform, dichloromethane

the structured packing material was not compatible with the treated VOCs. It was then replaced by spray chambers with dual-fluid nozzles. The disadvantage of the latter system was the high energy consumption. This system also caused an excessive water carryover to the biofilter. No demisters were installed, and over time the excess water that was carried over to the filter bed significantly increased the pressure drop of the first layer of packing. The spray nozzles for additional watering of the packing were clogged easily and needed cleaning or replacement on a yearly basis.

With experience, a regular water addition pattern was adopted for the biofilter. The first layer required the most water to compensate for the partially saturated inlet air stream and for the heat generated by the pollutant

Table 9.1 (continued)

Pollutants treated	Over the years, the total average concentration decreased from 500 mg m^{-3} to 180 mg m^{-3}, with less than 10 mg m^{-3} chlorinated. The pollutant emission is relatively stable over time (continuous 24 h per day)
Biofilter controls	Continuous monitoring (flow temperatures, pressure drop, inlet outlet concentrations, etc.) and logging into a PLC; moisture measurement by weighing the bed segments by load cells, manual control of the air flow through each section and of the water spray system
Biofilter design and acceptance criterion	Design based on pilot-test studies on-site, acceptance criterion: 80% removal of the volatile organics
Approximate investment costs	Approximately \$4,000,000; builder claims that it could be built now for about 50% less
Approximate treatment cost per 1000 m^3 off-gas treated	\$1.44 per 1000 m^3 off-gas treated (a detailed breakdown of the treatment costs is presented in Mildenberger and van Lith, 1996).
Performance	Initially 80% removal of organics, significant odor removal; see below

Figure 9.1 ARA-Rhein biofilter. (Courtesy of Monsanto Enviro-Chem; Chesterfield, MO.)

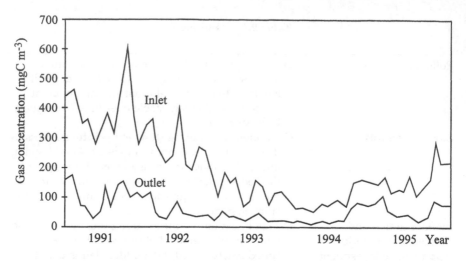

Figure 9.2   Average inlet and outlet concentrations for the period of 1991 to 1995. (Courtesy of Novartis, AG; Basel, Switzerland.)

biodegradation. The first layer of packing exhibited an increasing pressure drop over time, eventually reaching 50 cm water gauge. This was due to the structural degradation of the compost in the medium, and very fine particles preventing effective drainage of water. After 4 years of operation, the first layer was partially removed to keep the pressure drop within acceptable values. Also, because of excessive exposure to chlorinated compounds, the pH of the first layer dropped to 2 to 3 in less than 2 years. Because the pollutant loading was concurrently reduced due to upstream process improvement, this drop in pH and the removal of part of the first layer did not affect the overall pollutant elimination. The pH and the pressure drop over the second and third layer remained close to initial values for over 4 years.

9.2.2   Performance

Initially, the biofilter was exposed to short-term spikes of 10,000 ppmV of chlorinated compounds. After the shock loadings, it took 1 to 2 weeks to regain normal removal activity. The faster-than-expected exhaustion of limestone in the first layer of the biofilter was attributed to these spikes.

The average performance of the ARA-Rhein biofilter was about 15 g m^{-3} h^{-1} in the first two years at an average inlet concentration of 500 mg m^{-3}. The 80% removal acceptance criterion was met. Over the years the specific performance of the biofilter decreased (Figure 9.2), but the average inlet concentration decreased as well. This allowed the emissions to be kept under the allowable limits.

9.3 Odor control from flavor and fragrance manufacturing

Bush Boake Allen (BBA), Ltd., a subsidiary of Bush Boake Allen, Inc., which itself is a subsidiary of Union Cam, Inc., is a major manufacturer of flavors and fragrances. BBA has plants in the U.K. and throughout the world, and several use biofilters to control odor nuisances. Because of the numerous complaints received from as far away as 1.5 km, the Witham factory in the U.K. was obliged to schedule the manufacturing of many of its products for times when prevailing winds would not blow odorous emissions from its spray dryers over the nearby town. Extensive pilot plant testing was done at Witham by Monsanto Enviro-Chem (MEC)/ClairTech in 1990. A pilot biofilter successfully demonstrated that effective odor control could be achieved. It was not until 1995, however, that BBA decided to install a full-scale biofilter (Table 9.2), having tried unsuccessfully in the meantime to use a catalytic oxidizer to control the emissions. The latter failed because of fouling of the catalyst (Kampeter, 1998).

9.3.1 Design and operation comments

BBA initially required that the biofilter treat an air flow of 19,100 m³ h⁻¹ and be capable of reducing the level of the odor in the air discharged from the dryers from 200,000 odor units (OU) m⁻³ to less than 2000 OU m⁻³. BBA also specified that the biofilter should be capable of having its throughput increased by 50% at some time in the future without the need for major modification. A third requirement was that it should be constructed so that it could be moved to another site, if required. Accordingly, the filter was constructed from four mild steel modules to form a rectangular housing approximately 17.4 × 8.2 × 5.2 m with an integral humidifier in one corner. The internal surfaces were given an epoxy finish. To handle the initial air flow, 170 m³ of filter material were installed to make a bottom level 1.3 m deep. Four load cells were positioned beneath this layer. Sufficient space above this first layer of filter material was available to accommodate an upper level at a later date, to enable the capacity of the unit to be increased without an adverse effect on its performance. The belt-driven stainless steel fan (18.5 kW) was positioned so that the biofilter would operate under positive pressure. A standard PLC/PC control system was installed by MEC/ClairTech. MEC/ClairTech undertook the supply, installation, and commissioning of the biofilter system, but the concrete foundation and the ductwork from the spray dryers were supplied by others.

In 1996, BBA decided to have the capacity of the system increased by 58% to 30,150 m³ h⁻¹. MEC/ClairTech installed an upper layer of filter material 0.8 m deep and moved the load cells from beneath the lower layer of filter material and repositioned them to carry the upper layer. A larger capacity (35 kW) fan was installed and the control system software revised.

Table 9.2   Characteristics of the BBA-Witham Biofilter

Owner and location	Bush Boake Allen; Witham, UK
Builder	Monsanto Enviro-Chem/ClairTech BV; Woudenberg, The Netherlands
Type of air stream	Exhaust air from flavor, fragrance and fine chemicals manufacturer
Year of installation	1995 modified in 1996
Medium type and volume of medium	BIOTON® (50:50 compost and polystyrene spheres), 170 m³ initially; after the system was uprated the medium volume was 275 m³
Number of layers and height of medium	Originally one layer of 1-m height; after modification, a second layer of 0.8 m was added
Biofilter construction type	Closed biofilter made of a four mild steel modules with an epoxy coating inside; outside dimensions 17.4 × 8.2 × 5.2 m
Humidification	Packed bed tower
Air flow rate	Originally 19,100 m³ h⁻¹; 30,150 m³ h⁻¹ after modification; downward air flow
Empty bed residence time	32 seconds (unchanged)
Pressure drop	Biofilter: 500–600 Pa, including extensive ductwork: about 1200 Pa
Average bed temperature	20–25°C
Pollutants treated	Odors and fragrances, inlet concentration up to 200,000 OU m⁻³
Biofilter controls	Continuous monitoring (flow temperatures, pressure drop, etc.) and logging into a PLC; bed moisture monitoring using load cells
Biofilter design and acceptance criterion	Design should allow the air flow to be increased by 50% in the future or moved to another location; 99% odor removal was guaranteed at inlet concentration up to 200,000 OU m⁻³
Approximate investment costs	Initial costs about £300,000 ($480,000) upgrading costs were approximately £100,000 ($156,000)
Approximate treatment cost per 1000 m³ off-gas treated	Total treatment (including investment) costs were evaluated at £0.37 ($0.58) per 1,000 m³ off-gas treated; these costs reflect relatively high investment cost due to the specific configuration.
Performance	N/A

9.4   *Odor reduction from flavor manufacturing in a closed bed biofilter*

Bush Boake Allen (BBA), Ltd., has another biofilter successfully treating exhaust air from a flavor and fragrance manufacturing plant (Table 9.3; Figure 9.3).

Table 9.3   Characteristics of the BBA-Chicago Biofilter

Owner and location	Bush Boake Allen; Chicago, IL
Builder	PPC Biofilter; Longview, TX
Type of air stream	Exhaust air from food and flavor manufacturing
Year of installation	1995
Volume of medium and medium type	210 m³, patented mixture of inert and organic materials
Number of layers and height of medium	Two layers, 1.2 m high, total 2.4 m
Biofilter construction type	Modular construction, enclosed biofilter made of insulated epoxy coated steel, downflow mode
Humidification	Counter-current packed tower
Air flow rate	22,000 m³ h⁻¹
Empty bed residence time	35 seconds
Pressure drop	10 cm water gauge
Average bed temperature	24°C
Pollutants treated	Odors from flavor production and spray dryers: butter, grape strawberry, lemon, etc.; probable compounds: anthranilate, esters, ethers, aldehydes, variable concentrations; limited sampling showed about 2000 OU m⁻³
Biofilter controls	PLC-MMI (man/machine interface)
Biofilter design and acceptance criterion	No more odor complaints
Approximate investment costs	$375,000 delivered and installed
Approximate treatment cost per 1000 m³ off-gas treated	Yearly electricity costs: $11,500 for electricity, i.e., about $0.060 per 1000 m³ off-gas treated
Performance	95–99% odor removal

9.4.1   Design and operation comments

The system includes one independent counter-current packed tower humidifier equipped with a 7-hp recirculation pump. The air flow is provided by one 25-hp fan. From the humidifier, the gas flow tees off to two independent biofilters, allowing each unit to be serviced without interrupting treatment. Each biofilter is about 3.3 m wide by 13 m long and has two beds in series with a medium height of 1.2 m, for a total medium height of 2.4 m. The filter medium is expected to last 4 to 5 years. In the third year, the pressure drop over the medium was about 7.5 cm water gauge.

9.4.2   Performance

Removal efficiencies were evaluated by olfactometry once in 1996 (Standefer, 1998). They revealed that the butter smell had an inlet concentration of 1961

Figure 9.3 The BBA-Chicago biofilter.

OU m^{-3} and an outlet concentration of 39 OU m^{-3} (removal efficiency 98%); the strawberry odor had inlet and outlet values of 866 and 18 OU m^{-3}, respectively (removal efficiency 98%). Odor complaints dropped from 13 before the installation of the biofilter to zero after the biofilter was placed in service.

9.5 The Poughkeepsie biofilter: wastewater treatment odors

The City of Poughkeepsie wastewater treatment plant was receiving an increasing number of odor complaints. After considering various control technologies, it appeared that biofiltration would provide the most cost effective solution. An open-bed biofilter was installed (Table 9.4).

9.5.1 Design and operation comments

Various air sources were connected to the biofilter (Table 9.5). The main air duct work was made of stainless steel and had a diameter of 91 cm (36 inches). Upon startup, one of the dampers regulating the air flow of one of the source was inadvertently closed. This created a vacuum in the main air duct, causing the piping to collapse (Figure 9.4). The piping was replaced with heavier gauge stainless steel and a vacuum release valve was installed just upstream of the fan.

Table 9.4   Characteristics of the Poughkeepsie Biofilter

Owner and location	City of Poughkeepsie, NY
Designer	Webster Environmental Associates, Inc.; Pewee Valley, KY
Type of air stream	Exhaust air from a 10 million gallon per day wastewater treatment plant (various sources of air)
Year of installation	1996
Volume of medium and medium type	300 m³ of AllGro, a compost + bulking agent mixture by Wheelabrator, Inc.: 4 parts (by volume) hardwood and softwood chips; 1 part pine, spruce, or fir screened (>1 cm) bark; 1 part leaf finished compost screened to remove particles larger than 1.3 cm
	Medium is placed over a 38 cm thick layer of gravel, made of a bottom layer (2.5–5.1 cm gravel) which envelopes the air distribution system, and a finer (1–1.6 cm gravel) layer to prevent the medium to migrate and clog the air distribution system (Fuller, 1997)
Number of layers and height of medium	One layer of 91 cm over a 38-cm layer of gravel
Biofilter construction type	Open bed, above ground, 325 m² (3500 ft²) biofilter split into two cells (irregular shape)
Humidification	There is no prehumidification. Moisture (1 m³ h⁻¹ 12 hours per day) was provided by soaker hoses (1.6 cm diameter) at two layers within the bed, the first layer being at 15 cm from the biofilter top, the second at 60 cm from the top.
Air flow rate	25,800 m³ h⁻¹, upflow. The biofilter uses a 75-hp belt-driven fan
Empty bed residence time	41 seconds
Pressure drop	17.8 cm water gauge
Average bed temperature	Not determined
Pollutants treated	Mostly odors: hydrogen sulfide, dimethyl disulfide (DMDS) and methyl mercaptan.
Biofilter controls	Only a limited number of parameters are controlled: pressure drop, pH, and medium moisture content (by monthly grab sampling); soaker hoses on timer
Biofilter design and acceptance criterion	No more odor complaints
Approximate investment costs	$288,000
Approximate treatment cost per 1000 m³ off-gas treated	$0.20 per 1000 m³ off-gas treated (electricity and medium replacement) (Webster 1998)
Performance	98% removal of 10–25 ppm H₂S; odor complaints dropped to zero after correction of the short-circuiting

Table 9.5 Sources of Contaminated Air, Flow Rate, and Diameter of the Piping

Source	Flow rate (m^3 h^{-1})	Air duct diameter in cm (inches)	Type of emissions[a]
Gravity belt thickener	3400	30.5 (12)	10 h day^{-1} 1–2 ppm H$_2$S
Aerated grit chamber	4200	35.6 (14)	24 h day^{-1} 1–2 ppm H$_2$S
Primary clarifiers and low lift pumps	4200	35.6 (14)	24 h day^{-1} 2–5 ppm H$_2$S
Solid handling building	11,500	68.6 (27)	24 h day^{-1} 1–2 ppm H$_2$S
Sludge load out bin	2500	20.3 (8)	12 h day^{-1} 5–10 ppm H$_2$S + 1< ppm DMDS
Blower inlet/outlet	25,800	91.4 (35)	N/A
To biofilter inlet	2 times 12,900	Two pipes of 68.6 (27)	N/A

[a] See Webster (1998).

The biofilter was constructed as one large open bed, but air distribution was split into two separate systems that can be regulated by a damper in each of the headers (Figure 9.5). Air distribution was ensured by 6-inch (15.2-cm) diameter lateral PVC pipes spaced 122 cm apart in the lower gravel layer (Figure 9.6). The 6-inch pipe was perforated with two rows of 0.5-inch (1.3-cm) holes at a spacing of 6 inches (15.2 cm) and an angle of 45 degrees below the horizontal. No differential drilling or spacing of the holes was deemed necessary to ensure an even air flow. A 10-mil polyethylene liner was installed underneath the gravel to collect the drainage, which was returned to the front of the wastewater treatment plant (Webster, 1998).

9.5.2 Performance

After startup, suboptimal removal performance, in particular for DMDS, was observed. Smoke tests revealed that significant short-circuiting occurred on the sides of the biofilter (Webster, 1998). Note that the above-ground biofilter did not include any structural work to hold the medium in place and that the sides have a 3:1 slope (Figure 9.7). To correct this problem, wood chips were added to the side of the biofilter, and additional soaker hoses were added to maintain them at the appropriate moisture level. Smoke testing revealed that this corrective measure was successful. As pollutant removal went up, odor complaints stopped.

9.6 Soil biofilter to treat odors from a fabric softener facility

A factory near Chicago, IL, that produces fabric softener cloths was receiving odor complaints from neighbors. Although the chemical fragrance that masks

Figure 9.4 Close-up view of the three sections of stainless steel piping that collapsed under the vacuum. (Courtesy of R. Fuller.)

Figure 9.5 Overview of the main piping laid over a layer of coarse gravel. (Courtesy of R. Fuller).

Figure 9.6   One of the two distribution systems: perforated 6-inch (15-cm) pipes are connected to the 27-inch (69-cm) pipe via push joints. (Courtesy of R. Fuller.)

Figure 9.7   The almost completed biofilter. One can see the upper layer of soaker hoses on top of the biofilter bed. The biofilter was complete after another 15 cm of medium was placed over the soaker hoses. (Courtesy of R. Fuller.)

Table 9.6 Characteristics of the Fabric Softener Biofilter

Owner and location	Confidential; Aurora, IL
Builder	Bohn Biofilter Corp.; Tucson AZ
Type of air stream	Exhaust air from fabric softener factory
Year of installation	1991
Medium type and volume of medium	Soil, 353 m³; expected medium lifetime 30 years
Number of layers and height of medium	One layer of 0.75 m
Biofilter construction type	Above-ground, open bed, polygonal shape, 470 m²; enclosed construction of railroad ties; no thermal insulation; air distribution is through 4-inch (10-cm) ID slotted PVC pipe
Humidification	Fogger nozzle in main air pipe between blower and biofilter, additional water provided by rain and snowmelt
Air flow rate	8500 m³ h⁻¹, upflow
Empty bed residence time	3.5 minutes
Pressure drop	13-cm water gauge
Average bed temperature	Varying with seasons, estimated at 10°C in winter and 20°C in summer
Pollutants treated	Odors, mainly low concentrations of alcohols, ketones, and perhaps ethers; lemon fragrance probably also contained limonene, $C_{10}H_{16}$, which is only slowly biodegradable
Biofilter controls	No controls
Biofilter design and acceptance criterion	No detectable odors, no more odor complaints
Approximate investment costs	$65,000
Approximate treatment cost per 1000 m³ off-gas treated	Electricity cost mainly, i.e., $55 per year or $0.003 per 1000 m³ off-gas treated (0.375-kW blower, 40 hours per week, 7 cts kWh⁻¹)
Performance	Estimated odor removal of 99%, sustained over years

the odor of the fabric softener is pleasant and associated with cleanliness, high concentrations were considered a nuisance. An open-bed biofilter was installed (Table 9.6).

9.6.1 Design and operation comments

The compositions of the several fragrances are proprietary secrets of the factory's chemical supplier. They were certified to be non-toxic and were presumably non-bactericidal. The fragrances seemed to be mixtures of alcohols, ketones, and perhaps ethers, all of which are rapidly biodegradable.

The lemon fragrance probably also contained the hydrocarbon limonene, $C_{10}H_{16}$, which is more slowly biodegradable.

Because a full-scale system could not be designed on the basis of existing data, pilot tests were conducted. The tests were conducted outdoors during the months of December and January, in small biofilters with low thermal mass and no insulation around the sides. They showed that soil beds effectively removed the fabric softener odors even at low temperatures, eliminating the need to install costly temperature control.

The major odor sources — three impregnating roller machines — were then enclosed within the building so that 8500 m³ h⁻¹ (5000 cfm) could be drawn from the enclosures and blown into the biofilter. The air was humidified by a fogger nozzle in the pipe after the fan and within the building.

The soil on site was far too impermeable to use in a biofilter and too likely to form preferential flow paths. Suitable soil material was found within convenient trucking distance. The contractor was unsupervised and laid down the material with little regard for uniformity, so that the air flow distribution is non-uniform. The back-pressure was 12 cm water gauge (1.2 kPa). A lesson learned from this project was that supervising the construction of biofilter beds is recommended for optimum air flow and good biofilter performance. Building contractors rarely appreciate the need for careful placement of the media and for not compacting the bed.

9.6.2 Performance

Despite the non-uniformity of flow, odor removal by the biofilter was estimated at over 95% at the start. Much of the odor was from several small leaks which were obvious by moisture condensation in the cold weather. The limonene portion of the odor was stronger in the leaking air, indicating removal of other more biodegradable components. The odor removal improved to over 99% within a month as the leaks self-sealed by soil settling into the preferential flow paths that had developed initially during soil placement. Also, over time, endogenous microbes adapted to the fragrance chemicals. Since the biofilter was installed in 1991, the bed has received no maintenance. Rain and snowmelt make up any water deficit. Moisture control has not been a problem because of the relatively uniform annual precipitation at this location. Downpours have caused flooding problems. Odor control efficiency has not diminished, and the soil medium shows no sign of ever needing replenishment or replacement. Odor complaints have ended. Plant growth on the bed is very sparse. Ventilating the enclosures lowered the interior odor level dramatically, improving the quality of the work space.

Because the plant only operates during regular business hours, the biofilter blower is only turned on about 40 hours per week. This did not affect overall odor removal but allowed significant savings of electricity.

9.7 Small biofilters for gasoline vapor treatment at a soil-vapor extraction site

Four biofilters were installed at the Hayward, CA, soil-vapor extraction site to treat gasoline-contaminated air streams. Although this entire system is relatively small (total bed volume of 4.8 m^3), it can be considered full-scale because it is treating the entire exhaust from the soil-vapor operation (Table 9.7). Initially, four biofilter units were mounted in parallel. This arrangement was later changed to two parallel systems of two units in series, doubling the surface load while doubling the bed height.

9.7.1 Design and operation comments

The biofilters were made of cubic mild steel containers. All units had hooks to allow lifting with a crane and slots for fork-lift attachment. The cubic geometry was chosen to reduce investment costs; however, because of the back pressure of the carbon bed downstream of the biofilters, significant side-wall deflection occurred. When the sides moved outwards, they separated from the medium and caused air short-circuiting. A general rule is that cylindrical geometry is better suited for resisting side-wall deflection, especially if a significant back pressure is expected downstream of the biofilter. Operating the biofilter under negative pressure with a blower in a vacuum (induced draft) mode could have solved the air channeling problem caused by the side-wall deflection.

The humidification chamber was made of a tank recovered from a previous project. It was oversized and included two multi-nozzle Fog-Jet manifolds (Spray Systems; Wheaton, IL) in the upper center region of the 91.4-cm (diameter) × 183-cm (length) cylindrical humidification chamber which dispensed a 60-mm spray countercurrent to the gas stream. The contaminated gas stream entered the humidifier tangentially from a lower port. Volumetric flow rate of the spray was adjustable from 0 to 57 L h^{-1}. Water not entrained was recirculated. Make-up water was introduced into the reservoir automatically using a float valve with a high-level cut-off switch.

Maintaining proper bed moisture proved to be the major challenge in these biofilters. The temperature-compensated, resistance-type soil moisture probes were only able to provide a qualitative tracking of the medium moisture in the vicinity of the probes. Bed drying was diagnosed only when a significant decline of pollutant removal was noticed after 100 days of operation. For bed moisture mitigation, the air flow rate was momentarily reduced, and for one unit, the medium was removed and thoroughly mixed. All units received 200 L of water, but most of this water was recovered in the drainage within one hour. This illustrates the difficulty of rehydrating a partially dried bed. After this event, supplemental water was regularly added to the top of the biofilters. This was sufficient to maintain proper moisture

Table 9.7 Characteristics of the University of California, Davis/
Environmental Resolutions Biofilter

Owner and location	EXXON service station; Hayward, CA
Builder	University of California, Davis, and Environmental Resolution, Inc.
Type of air stream	Exhaust from soil vapor extraction (SVE), site contaminated with gasoline
Year of installation	1995
Medium type and volume of medium	Mixture of sewage and wood products compost and perlite (50:50); packing was amended with 1 eq $CaCO_3$ $kg^{-1}_{packing}$ as crushed oyster shell, and 8 kg m^{-3}_{medium} KNO_3; two of the four units received 25 kg m^{-3} of a controlled-release nutrient (18-06-12); each unit had a bed volume of approximately 1.2 m^3
Number of layers and height of medium	One layer of 0.91 m per biofilter (upflow mode); a layer of 5.1 cm crushed cedar bark was placed between the inlet support grid and the compost-perlite medium; for the first 8 months, four units were operated in parallel, and after that the system was reconfigured to two systems of two units in series
Biofilter construction type	Mild steel container; emphasis placed on minimizing construction costs; biofilters were 1.2 × 1.2 × 1.2 m, with a lower inlet plenum of 15.2 cm and upper plenum of 10.2 cm; drainage ports were located at the bottom of each unit; two 100-kg GAC canisters in series were installed downstream of the biofilters to ensure compliance
Humidification	One counter current spray tower (0.9-m diameter, 1.8 m in height) with water recycle for the four biofilters; water was not heated; average water droplet diameter was 60 μm; no demister was installed after the humidification
Air flow rate	13–32 m^3 h^{-1} (upflow) each when the four units were mounted in parallel (a bed moisture mitigation period occurred at 4 m^3 h^{-1}) 44–47 m^3 h^{-1} (upflow) each when the biofilters were configured as two systems of two units in series

conditions, but according to Wright et al. (1997), optimum conditions would have probably resulted in better overall performance. A consequence of the mode of addition of water by flooding the top of the biofilter is that it separated the compost and the perlite in the top few centimeters and resulted in an increased pressure drop. As a whole, the problems experienced with moisture control suggested that such biofilters should include a number of

Table 9.7 **(continued)**

Empty bed residence time	1.6–3 min (10 min during bed moisture mitigation)
Pressure drop	Initially less than 1 cm of water column, after 8 months ranging from 6.4–15 cm
Average bed temperature	Variable with seasonal changes and with sun exposure: 15–36°C
Pollutants treated	Gasoline tentatively identified as methyl alkanes and C_6–C_9 cycloalkanes, and BTEX (mainly xylene isomers); BTEX amounted for about 20% of total petroleum hydrocarbons (TPH); TPH inlet concentration was initially up to 2.7 g m^{-3}, was diluted to 0.8-1.4 g m^{-3} for more than 200 days, and finally declined to about 0.4 g m^{-3}
Biofilter controls	Two of the biofilter units had soil moisture sensing probes. The control and monitoring was essentially manual with grab samples taken at regular intervals. The humidification chambers have an automatic water supply with low-liquid level pump shut off.
Biofilter design and acceptance criterion	Unknown
Approximate investment costs	$12,000 total for the four biofilters, blower and humidification system
Approximate treatment cost per 1,000 m³ off-gas treated	Electricity: $150 to $300 per month for the four biofilters — assuming a 2-year unit life (and project life), a 10% interest rate, and total air flow of 170 m³ h^{-1}: operating costs $1.21–$2.42/1000 m³; investment costs $4.23/1000 m³; total treatment costs: $5.44–$6.65/1000 m³
Performance	Removal of TPH of 90% could be maintained for a significant period of time (inlet concentration of 0.4 g m^{-3}, EBRT of 1.8 min); over 400 days, cumulative removal shows approximately 60% removal of TPHs, with about 75% removal of BTEX

access ports along the bed length to allow for grab samples and moisture checks. It was recommended that subsequent systems have a simple method of adding water to the biofilters such as a soaker hose placed at the top of the unit for water addition once or twice per week depending on the air flow rate and inlet air relative moisture.

9.7.2 Performance

The inlet and outlet TPH concentrations and removal percentage of the biofilter units are reported in Figure 9.8, while the chronology of the events

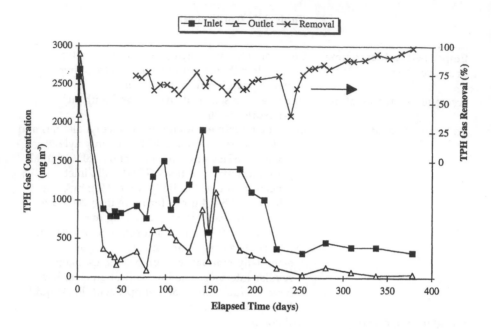

Figure 9.8 Biofilter inlet and outlet concentrations and TPH removal percentage as a function of time. (Adapted from Wright,W.F. et al., *J. Environ. Eng.*, 123(6), 547, 1997. With permission.)

is reported in Table 9.8. Initially, the biofilters were inoculated with gasoline-degrading enrichment cultures. Even so, it took more than 22 days for the biofilters to become effective. The reasons probably were the high initial pollutant concentrations (2.7 g m⁻³) which were detrimental to the establishment of an active biofilm, the repeated shutdowns of the experimental units, and the fact that after inoculation, biofilter operation did not start immediately. It was recommended that a biofilter unexposed to gasoline vapors for an extended period of time (e.g., after an extended shutdown or prior to actual startup) be supplied with an artificial supply of gasoline to maintain optimum bacterial activity. After the original acclimation phase, the biofilters were able to degrade 70 to 90% of the TPHs depending on the conditions. BTEX (mainly xylene isomers) removal was typically greater than 90%. GC and GC/MS of gas samples demonstrated that methyl-substituted alkanes and cycloalkanes tended to be more recalcitrant.

The temperature of the biofilters varied with the seasons and with the sun exposure. During the winter time, the temperature dropped as low as 11°C. Because of the cold weather, the mass loading rates to the biofilters decreased in winter, so that direct comparison of biofilter performance is unfortunately not possible. In general, low temperature did not seem to

Table 9.8 Chronology of Events in the Biofilters

Days	Characteristic operation	EBRT (min.)	System shutdowns (total duration in days)	Pressure drop across biofilter (cm water gauge)
0–22	Acclimation, negligible removals, frequent shutdowns	1.6	1 (6)	<0.5
23–115	Moderate inlet concentrations, significant removal rates and subsequent onset of bed drying	2.7	1 (0.8)	0.5–0.8
116–137	Bed moisture mitigation, discontinuous operation	10.1	1 (4)	<0.5
138–252	Recovery of biofilter performance	2.3	4 (6.2)	1.0
253–378	Change in configuration (reactors coupled in series)	1.8[a]	2 (1.6)	6.4–15

[a] Total EBRT for the two biofilters in series.

affect pollutant removal significantly, but the placement of the units — two in the shade and two in the sun — had an impact on pollutant elimination and on drying rates.

The cumulative TPH and BTEX removal for the first year of treatment in the four biofilters is reported in Figure 9.9. The data show that significant savings in granular activated carbon (GAC) were achieved. These were evaluated by Environmental Resolutions, Inc., for a 2.8-year period as follows (ERI, 1998):

Mass of hydrocarbons removed by
the system during the period of
3/15/95 to 1/21/98: .... 2000 kg

Mass of GAC required to abate
2000 kg of gasoline hydrocarbon
according to GAC manufacturer
(15% loading on GAC) .... 13,390 kg

Mass of GAC actually used in the
same period .... 5220 kg

Savings in GAC (mass) .... 8170 kg

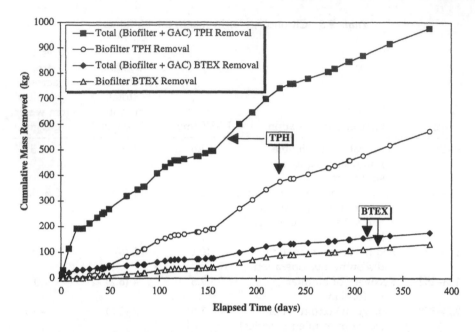

Figure 9.9 Cumulative TPH and BTEX mass removal over treatment time course. (Adapted from Wright,W.F. et al., *J. Environ. Eng.*, 123(6), 547, 1997. With permission.)

Savings in GAC (cost of GAC,
$4.4 kg^{-1} including service company's
delivery and removal) $36,000

Estimated savings over 2.8 years using
biofilter-GAC combination vs. GAC only
(2.8 years of biofilter use including
biofilter purchase of ~$20,400) $15,600

Wright et al. (1997) speculate further, that for similar treatment, if correct water management is implemented and inlet TPH concentrations are maintained at a maximum of 0.4 g m^{-3}, GAC units downstream of the biofilters may not be necessary, depending on local regulation.

9.8   Treatment of VOC mixtures from exhaust air in the wood industry

A very large air stream from a particleboard press vent is treated in an enclosed biofilter (Figure 9.10; Table 9.9).

Figure 9.10   The Weyerhaeuser biofilter. (Courtesy of PPC Biofilter; Longview, TX.)

9.8.1   Design and operation comments

Reactor control is ensured by a PLC-MMI (programmable logic controller with man/machine interface) providing a real-time display and history of key operating parameters. It includes eight differential pressure transmitters to measure pressure drop over each independent humidifier and each independent biofilter bed, six relative humidity transmitters, 20 thermocouples, 16 load cells, four water meters to record the daily application of water to the filter medium, inlet and outlet concentrations, and relays to actuate the four humidifier recirculation pumps and the eight fans. Alarm conditions are set for differential pressure and temperature with bypass protection when needed.

The pretreatment humidifier (900 m^3 h^{-1} recirculation rate) evaporates about 53 m^3 of water per day. Approximately the same amount is applied directly to the biofilter using overbed sprays to maintain a constant moisture content. Filter medium leachate is drained back to the pretreatment humidifier, bringing biologically active water and some nutrients to the pretreatment system for the removal of soluble compounds such as alcohols and formaldehyde.

The particulate loading is about 8.2 kg h^{-1}, which is reduced to about 1.8 kg h^{-1} in the pretreatment humidification system. Suspended solids are maintained at a reasonable value in the recycle water by a blow-down of 680 L h^{-1} to a centrifuge.

Table 9.9   Characteristics of the Weyerhaeuser Biofilter

Owner and location	Weyerhaeuser; Adel, GA
Builder	PPC Biofilter; Longview, TX
Type of air stream	Particleboard dryer
Year of installation	1995
Volume of medium and medium type	2500 m^3, patented mixture of inert and organic materials
Number of layers and height of medium	One layer, 1.5 m high
Biofilter construction type	Concrete, down-flow operation, four independent reactors
Humidification	2-stage, high-velocity quench and packed tower
Air flow rate	245,000 m^3 h^{-1}
Empty bed residence time	37 seconds
Pressure drop	5 cm of water gauge
Average bed temperature	21–38°C
Pollutants treated	Formaldehyde, alpha- and beta-pinene, about 0.2 g m^{-3}
Biofilter controls	PLC-MMI
Biofilter design and acceptance criterion	Minimum 80% removal
Approximate investment costs	Confidential; similar system would cost about $4,300,000 delivered and installed (Standefer, 1998)
Approximate treatment cost per 1,000 m^3 off-gas treated	Yearly operating costs: $83,000 for electricity, $7000 for water, $9000 for maintenance; this corresponds to $0.046 per 1000 m^3 off-gas treated
Performance	85–99% removal, depending on temperature

9.8.2   Performance

The average bed temperatures fluctuate with the temperature of the gas stream and the ambient conditions. In this particular application, these fluctuations strongly influence the pollutant removal efficiency. At 95°F (35°C), 95% removal is achieved, whereas at 85°F (29°C) the removal of the alpha- and beta-pinene is approximately 85%, i.e., an approximate increase of 1% per degree Fahrenheit (Standefer, 1998).

9.9   *Control of VOCs from ink-drying operations*

Serigraph, Inc., prints plastic substrates such as speedometers, tachometers, and appliance and computer parts. The printed products are dried in a natural gas-fired drying oven. The special inks contain various VOCs which are volatilized during drying, and the exhaust air requires treatment. The air pollution control (APC) permit requirements for Serigraph are based on total tons per year of VOC emission and not on percentage removal basis. Hence,

Figure 9.11 The Serigraph biofilter. (Courtesy of PPC Biofilter; Longview, TX.)

Serigraph calculated that for its growth and future expansion, the chosen air pollution control technology should achieve a minimum of 80% VOC removal. A review of possible control technologies showed that biofiltration was much more cost effective than incineration and other techniques. The treatment costs in a biofilter were about $5900 per ton of VOC removed, and as high as $20,000 per ton of VOC if an incinerator-concentrator combination was to be used. A 400-m² biofilter was installed in 1997 (Figure 9.11; Table 9.10).

9.9.1 Design and operation comments

Air pretreatment is achieved in two independent humidifiers. Each one is 9 m long by 1.5 m wide by 5 m tall and includes a 1.5-m high structured packing for countercurrent gas-liquid contact. A spray header above the packing delivers 200 m³ h⁻¹ of recirculation water to the packing. The air comes into the humidifier at approximately 65°C and 30% relative humidity and leaves at about 27°C and 97% relative humidity. Demisters are installed at the exit of the humidifier to prevent excess carryover of water droplets to the biofilter bed. Additional moisture is provided by six independent overbed spray systems with filters and strainers for fine mist nozzles. Each section is controlled by a valve actuated by the PLC timer. The PLC adjusts the readings of six individual load cells located under the air distribution grids for

Table 9.10   Characteristics of the Serigraph, Inc., Biofilter

Owner and location	Serigraph, Inc.; West Bend, WI
Builder	PPC Biofilter; Longview, TX
Type of air stream	Exhaust from a silkscreen printing dryer
Year of installation	1997
Volume of medium and medium type	600 m³, patented mixture of inert and organic materials
Number of layers and height of medium	One layer, 1.5 m high
Biofilter construction type	Concrete, two independent reactors, down-flow mode
Humidification	Counter-current packed tower and additional overbed spray
Air flow rate	76,000 m³ h⁻¹
Empty bed residence time	28 seconds
Pressure drop	5 cm water gauge
Average bed temperature	29°C
Pollutants treated	Alcohols (5–8%), ethers (10–15%), esters (35–40%), ketones(13–15%), aliphatic (8–11%) and aromatic (12–15%) hydrocarbons, at least 22 individual compounds; average total inlet concentration about 0.1 g m⁻³; significant concentration of methane remaining from the burners
Biofilter controls	Sophisticated PLC-MMI (man/machine interface); large number of operating parameters such as air temperature, moisture, pressure, etc. continuously monitored; load cells used to monitor the moisture content of the medium
Biofilter design and acceptance criterion	>80% removal
Approximate investment costs	Confidential; a similar system would cost about one million dollars delivered and installed
Approximate treatment cost per 1000 m³ off-gas treated	Yearly operating costs: $43,000 for electricity and $8000 for water; this corresponds to $0.065 per 1000 m³ off-gas treated
Performance	85–95% removal of non-methane hydrocarbons; as the inlet concentration goes below 15 ppm, the removal efficiency decreases due to diffusion limitation

differential pressure and calculates the operating biofilter bed weight. Depending on this value, the PLC activates the overbed sprinklers in addition to the regularly timed watering cycles.

The biofilter is composed of two independent units. The biofilter is constructed of 20-cm thick concrete, with a 20-cm thick hollow core roof and 10 cm of insulation to protect against excessive condensation on the media

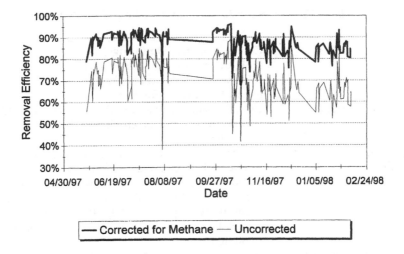

Figure 9.12 History of VOC elimination in the Serigraph biofilter. During this period, the inlet concentrations were continuously between 30 and 70 ppm. (Courtesy of PPC Biofilter; Longview, TX.)

during the winter months. One unit is capable of handling the process exhaust, while the second reactor is off-line during maintenance or medium replacement so that production can run continuously. The biofilter includes two fans with variable frequency drives for energy efficient air flow control.

9.9.2 Performance

Typical removal performance is reported in Figure 9.12. It should be noted that the actual reactor performance was probably higher. The reason is that oxygenated compounds present in the biofilter inlet which have a lower response factor on the flame ionization detector (FID) were better removed in the biofilter than other compounds with higher FID response factors. Hence, the true concentration of VOC in the biofilter outlet was somewhat lower than actually measured by the FID.

9.10 Removal of high concentrations of ethanol from a foundry off-gas

A two-level, open biofilter treated 17,000 m³ h⁻¹ (10,000 cfm) of off-gas from the mold production at an investment foundry in Los Angeles from 1993 to 1996 (Table 9.11). Depending on the production rates, the off-gas contained often high (up to 3 g m⁻¹) and strongly fluctuating concentrations of VOCs, predominantly ethanol. Under these conditions, the biggest challenge was to maintain optimum bed moisture (Leson 1993; Leson et al., 1995).

Table 9.11   Characteristics of the Cast Alloys, Inc., Biofilter

Owner and location	Cast Alloys, Inc.; Northridge, CA
Builder	BioFiltration, Inc.; U.S. (licensee for G+E Umwelttechnik GmbH; Aachen, Germany)
Type of air stream	Exhaust air from a foundry
Year of installation	1993; the biofilter was shut down in spring of 1996 after process change eliminated the need for the biofilter
Volume of medium and medium type	315 m³ of a mixture of sewage sludge compost and pine bark medium; approximately 20 kg m⁻³ of pelletized calcium carbonate were added as pH buffer; initial pH of the medium was 7.2, and density was approximately 600 kg m⁻³; a 15-cm thick layer of bark was placed over the biofilter medium
Number of layers and height of medium	Two layers, 1.5 m high (in parallel)
Biofilter construction type	Open-bed biofilter with two layers of medium on top of each other but operating in parallel; outside dimensions of the concrete structure: 16.8 × 6.7 × 7.3 m; total bed area 210 m²
Humidification	Spray tower humidifier pre-conditioning and additional water sprinklers (24 spray nozzles per bed) on top of each bed regulated by a timer; for pH mitigation, a sodium bicarbonate solution of 0.5% by weight was periodically added into the irrigation water; drainage due to excess irrigation and condensation water in air ducting was directed to the sewer

9.10.1 *Design and operation comments*

The system was designed to remove on average 90% of a daily load of about 530 kg of ethanol. Design specifications included that peak ethanol concentration should not exceed 3.5 g m⁻³ for more than 10 minutes. An average elimination capacity of approximately 80 g m⁻³ h⁻¹ was assumed, so that a residence time of approximately 60 to 70 seconds was calculated. The biofilter design included an electric 40-hp centrifugal blower with an adjustable frequency drive (Figure 9.13). Although this was more expensive than dampers, it allowed energy-efficient control of the air flow (Leson, 1993). The main air duct was made of 30-inch (76-cm) diameter galvanized steel. Off-gas preconditioning included humidification to near saturation in a counter-current spray tower, which also served to remove silica dust entrained in the off-gas. A demister was installed between the humidifier and the blower to

Table 9.11 (continued)

Air flow rate	17,000 m³ h⁻¹, upflow
Empty bed residence time	60–70 seconds
Pressure drop	Initially, pressure drop was less than 4 cm water gauge; over time, it increased to more than 30 cm water gauge; it remained below 5 cm water gauge after a change in the medium
Average bed temperature	23–25°C
Pollutants treated	Ethanol, highly variable concentrations typically between 0.5 and 2 g m⁻³, sometimes up to 3 g m⁻³; traces of methanol, acetone, and about 1–5% poorly biodegradable chlorofluorocarbons; typically 24 hours a day, 5–7 days a week
Biofilter controls	Process monitoring included sensors for air relative humidity, temperature, off-gas velocity, pressure, meters for water consumption, and timer for additional humidification
Biofilter design and acceptance criterion	90% removal of reactive organic compounds (ethanol + traces of other volatiles)
Approximate investment costs	$550,000 including auxiliary equipment (see Leson, 1993)
Approximate treatment cost per 1000 m³ off-gas treated	Energy costs evaluated at $0.63 to $2.43 per hour; medium costs $20,000 per year (3-year lifetime, $200 m⁻³ installed); operating costs: $0.17–$0.28 per 1000 m³ off-gas treated; total treatment costs (10-year lifetime, 8% interest): $0.72–$0.83 per 1000 m³ off-gas treated
Performance	80–90% removal of total VOC

prevent excess carryover of water to the biofilter. Water consumption for the biofilter was evaluated and amounted to about 30 to 60 L h⁻¹ for the humidifier and 60 to 120 L h⁻¹ for the additional watering of the bed. The water discharge from the humidifier was 15 to 30 L h⁻¹ to keep the total suspended solids and salt concentrations at a reasonable values, and 30 to 60 L h⁻¹ from condensation and bed drainage. Significant bed watering — and consequently drainage from the bed — was required because of non-homogenous sprinkling over the biofilter and because of rapid percolation of the moisture through the bed. Per 1000 m³ off-gas treated, water consumption and discharge values corresponded to 5.7 to 11.4 L of fresh water and 2.8 to 5.6 L of wastewater, respectively (Leson, 1993).

The air distribution used patented BIKOVENT® (Braintech GmbH; Herzogenrath, Germany) concrete blocks allowing homogenous air distribution while allowing for drainage of excess water. Medium installation in

Figure 9.13 Biofilter, blower, and air ducting. At the time of the photograph, the biofilter was covered with a temporary steel-tarp structure to allow for representative sampling of the exhaust air. (Courtesy of G. Leson.)

the lower bed was achieved with a front-end loader through the side windows. Installation of the upper bed was from the top of the concrete structure using a crane and a funnel of the type normally used for concrete work. It took about 4 days and a crew of four persons to install the medium in the biofilter.

9.10.2 Performance

Initially, the biofilter achieved high removal efficiencies (>95%) and elimination capacities (up to 180 g $m^{-3} h^{-1}$) for total VOC; however, at these high concentrations, exothermic effects became significant. At inlet wet bulb temperatures of about 20°C, temperature increases across the bed of more than 5°C and moisture losses from the medium of more than 500 g $m^{-3} h^{-1}$ were common. Although a timer based, semi-automatic sprinkler system controlled bed spraying, maintenance of the proper bed moisture at these evaporation rates was time consuming and did not prevent repeated drying, particularly in the bottom zones. The considerable need for irrigation, typically 2 cm day^{-1}, also caused flush-out of fine particles which were present in excess in the original bark and sewage sludge compost filter material. The fine particles partially clogged the air distribution system and combined with the bed compaction (15 to 25 cm) resulted in an increase in bed pressure drop to more than 3000 Pa (30 cm water gauge). After 2 years of operation, corrective action was necessary, and the medium was replaced with a wood-chips-based medium that was less susceptible to compaction. After medium replacement, the pressure drop remained below 500 Pa for the remainder of the system's operation (Leson et al., 1995).

At the initially high inlet concentrations and VOC loadings, the biofilter also generated and released products of incomplete biooxidation, primarily acetic acid, causing odor problems, gradual acidification of the media, and corrosion of the galvanized ductwork. Process changes resulting in load reduction, and the periodical flushing with 0.5% sodium-bicarbonate solution eventually eliminated both cause and undesirable effects. The typical frequency for bicarbonate wash was about 180 kg of $NaHCO_3$ every 4 to 6 weeks (Leson, 1993). The acidification phenomenon was subsequently discussed by Devinny and Hodge (1995) and was attributed to the limited transfer of oxygen into the biofilm, causing the release of biodegradation intermediates at high VOC loadings.

This project demonstrated that biofilters achieve excellent removal of high loads of ethanol; however, it also pointed out some of the fundamental limitations of the technology when applied to off-gases containing high concentrations of water soluble VOCs. In particular, it demonstrated that the high rates of evaporation caused by the system's activity demanded the use of a fully automatic moisture-control system and operation in the down-flow mode.

Table 9.12   Characteristics of the Mercedes-Benz Biofilter

Owner and location	Daimler-Benz AG; Untertuerkheim plant, Germany
Designer and builder	Daimler-Benz AG; Untertuerkheim plant, Germany
Type of air stream	Exhaust air from a foundry (iron and light metals)
Year of installation	1992
Medium type and volume of medium	Three parallel modules with different media: (1) wood chips, approximately 3 cm in length, (2) peat and humus, (3) compost and expanded clay; total volume: 850 m³
Number of layers and height and of medium	One layer, 1.5 m high
Biofilter construction type	Open-bed biofilter with three beds in parallel, total bed surface of 567 m²
Humidification	Venturi humidifier (pressure drop about 4500 Pa), additional sprinkler on top of the bed
Air flow rate	80,000 m³ h⁻¹, upflow
Empty bed residence time	38 seconds
Pressure drop	Wood chips medium: mostly 100 Pa (up to 800 Pa) Peat and humus medium: 100–300 Pa (up to >2000 Pa) Compost and expanded clay medium: 250–500 Pa (up to >2,000 Pa)
Average bed temperature	20°C (average air temperature)
Pollutants treated	Mixtures of aromatics (toluene, xylenes, phenol), formaldehyde, odorous compounds; total concentration of organic carbon about 150 mg C m⁻³; average aerosol concentration 15–20 mg m⁻³
Biofilter controls	On-line measurement of inlet and outlet pollutant concentration
Biofilter design and acceptance criterion	Based on pilot tests, removal of 60% of organic compounds and 80% of odors was expected
Approximate investment costs	2,800,000 DM ($1,708,000)
Approximate treatment cost per 1000 m³ off-gas treated	1.0 DM per 1000 m³ off-gas treated ($0.61)
Performance	35–70% organic carbon, 50-80% odor removal; fluctuations observed

9.11 Large open-bed biofilters for the removal of VOCs and odors

In the early 1990s, Mercedes-Benz of Germany experienced repeated odor complaints from populations neighboring its foundries. Evaluation of air pollution control technologies showed that both adsorption and thermal oxidation would be too expensive. A series of biofilter pilot and technical-scale tests (500 to 2000 m³ h⁻¹) was performed, resulting in design and construction of five full-scale biofilters with a total capacity of 330,000 m³ h⁻¹. Described in Table 9.12 is one biofilter treating 80,000 m³ h⁻¹ of air at a foundry in Stuttgart, Germany.

9.11.1 Design and operation comments

Because of the presence of aerosols, a venturi scrubber was used for the humidification of the waste air. It was estimated that the electricity costs could be reduced by about 65% if particles and aerosol concentrations were lower and a normal spray tower humidifier could be used. The venturi scrubber was not fully effective in capturing aerosols and particles. Within 3 months of biofilter startup, the pressure over the filter bed increased from below 500 Pa to more than 2000 Pa. The reason was that a 1- to 2-cm thick layer of coconut fibers had been placed between the grid holding the medium and the biofilter medium and it effectively captured aerosols and rapidly clogged. The problem was partially solved when the layer of coconut fiber was removed. It was suggested that a down-flow biofilter could probably be advantageously operated with such a layer or an aerosol filter laid on top of the bed. This layer would be periodically replaced as needed.

Peat with heather, chopped wood, and compost with expanded clay media were tested. Over the long run, only the wood-chip medium proved cost effective, but it required continuous supply of mineral nutrients. The peat-heather medium had to be replaced within a year of operation because of excessive pressure drop. Overall, the wood-chip medium exhibited slightly lower pollutant removal than the other media, but it was preferred over the others because the capital and treatment cost would otherwise have been too high. Interestingly, compaction of the peat/heather and of the compost/expanded clay medium was increased by vibrations from the foundry building (Paul and Nisi, 1996).

In order to evaluate the disposal of the used medium, it was analyzed for various metals. The results are shown in Table 9.13 and compared to the maximum allowable concentrations in domestic waste compost. Over time, it was observed that concentrations of some metals (lead, cadmium, and zinc) in the medium could reach levels at which disposal could become an issue whereas others remained well below the maximum allowable concentrations.

The investment and treatment costs for the 80,000 m³ h⁻¹ waste air flow were as follows (in 1994, 1 DM = $0.61):

- Investment costs .... 2,800,000 DM ($1,708,000)
  With 8% interest over 10 years, this represents 504,000 DM ($307,440) per year
- Operating costs (yearly)
  Electricity .... 190,000 DM ($115,900)
  Water .... 25,000 DM ($15,250)
  Biofilter medium .... 45,000 DM ($27,450)
  Total .... 260,000 DM ($158,600)
- Total treatment costs (yearly): 764,000 DM ($466,040) or 1.090 DM per 1000 m³ air treated ($0.665)

Table 9.13 Concentration of Various Metals in
Biofilter Medium After the Pilot Tests

Element	Concentration (mg kg^{-1} dry medium)	Maximum allowable concentration[a] (mg kg^{-1} dry medium)
Lead	122	150
Cadmium	1.8	2
Chrome (total)	55	100
Copper	55	100
Nickel	17	50
Zinc	589[b]	400
Mercury	0.4	1.5

[a] According to RAL 5620, i.e., German norm for household compost.

[b] The results of subsequent analyses for the medium of full-scale units showed lower zinc concentrations in the range of 180 mg kg^{-1} medium even after 6 years of use.

Note: Medium moisture content was 65.5%.

Source: From Paul, H., in *Biologische Abgasreinigung*, VDI Berichte 1104, VDI Verlag GmbH, Duesseldorf, Germany, 1994, p. 355. With permission.

Note that this calculation shows that capital costs represent a non-negligible part of the overall treatment costs.

Over the years, a monitoring protocol was developed for the biofilters at Mercedes-Benz. It is summarized in Table 9.14.

9.11.2 Performance

Typical removal performance by the chopped wood medium was 50 to 80% of odors at inlet concentrations of 500 to 1000 OU m^{-3} and 35 to 70% of organics at inlet concentrations of 100 to 150 mg C m^{-3} at volumetric loadings of 150 to 225 m^3 m^{-3} h^{-1} (Paul and Nisi, 1996). Performance fluctuated and was considerably influenced by factors such as volumetric loading, crude gas concentration, and nutrient supply. Low temperatures were detrimental to the removal of VOCs. Typically, VOC removal increased linearly from about 30% at 13°C to about 80% at temperatures between 26 and 29°C. Heating the air stream to obtain higher pollutant elimination was never considered because of the high cost.

After 2 years of continuous operation, preferential air paths developed. Air velocity measurements at the surface of the biofilter showed velocity differences up to a factor of 6. Air short-circuiting coincided with a reduction of the biofilter performance. The medium was mixed and the original pollutant removal was fully recovered.

Table 9.14 Proposed Monitoring Schedule for the Mercedes-Benz Biofilter

Frequency	Parameter
Continuous	Inlet and outlet pollutant concentrations
Three times per week	Air flow rate[a]
	Pressure drop[a]
	Inlet air relative humidity[a]
	Inlet air temperature[a]
	Bed temperature[a]
	Rain
	Overbed sprinkling water flow
	Spray nozzle inspection
Weekly	Visual examination of the water in the humidifier
Biweekly	Visual examination of the top of the biofilter bed for dry zones, cracks and irregularities
Monthly	Moisture content of the medium (e.g., at two different depths)
Quarterly	Distribution of the air velocity over the filter bed surface (e.g., 20 measurements)
	COD, BOD_5, pH, and conductivity of the biofilter bed leachate
	Medium respiration activity and medium total heterotrophic count
	Olfactometric assessment of the biofilter performance
Yearly	Organic, inorganic, and ash content of the biofilter medium
	Analysis of the medium for heavy metals
As required, after process change or abnormal conditions	COD, BOD, TOC, conductivity, pH, GC/MS analysis of the water in the humidifier
	Analysis of biofilter inlet and outlet air samples by GC, possibly GC/MS

[a] These parameters will be monitored on-line in the near future.

To date, the chopped wood medium has not been changed. Sustained performance is ensured by a periodic monitoring of the biofilter (see Table 9.14) and a continuous supply of nutrients.

9.12 High-concentration, low-flow biofilter for VOC treatment

A coating company uses high-grade catalyzed paints for application to a variety of manufactured parts. The process produces many cleaning rags, paint cans, and still bottoms that must be disposed of. They use a forced-air dryer and biofilter combination manufactured by Bio-Reaction Industries to evaporate and biodegrade the solvents. The biofilter is designed

Table 9.15 Characteristics of the Rag Dryer Biofilter

Builder	Bio-Reaction Industries; Tualatin, OR
Type of air stream	Vapors from rag drying, paint sludges, still bottoms
Medium type	Proprietary compost-inorganic mix
Number of layers and height of medium	Two layers; one tray 0.3 m, one tray 0.15 m in height
Biofilter construction type	Enclosed tank, 2-m total reactor height, bed made of two trays of medium; system includes hot water circulation for temperature control and a separate process control system; also includes a drying chamber, e.g., for rags and other VOC waste disposal or maybe connected to a waste air collection system, e.g., aerosol can puncturing system
Humidification	Sprinkling of each tray
Air flow rate	1.7 to 3.4 m^3 h^{-1}, upflow
Empty bed residence time	12 min in first tray, 6 min in second tray
Pressure drop	<1 cm water gauge
Average bed temperature	26–38°C
Pollutants treated	Solvents — primarily toluene, but also methyl isobutyl ketone, methyl ethyl ketone, xylene, acetone, alcohols; highly variable concentrations with surges up to 20,000 ppm
Biofilter controls	Sprinkling of each tray controlled by a timer; automatic nutrient addition
Approximate investment cost	Unknown
Approximate treatment cost per 1000 m^3 off-gas treated	Electricity costs are about $1.8 per day, maintenance requires 15 min. per week (see Stewart and Thom, 1996); vendor reported $13.60 per day over 10 years, $9.00 per day for newer designs
Performance	>95%

for exceptionally low flows of 1.7 to 3.4 m^3 h^{-1} (1 to 2 scfm) and high, variable contaminant concentrations (Table 9.15).

9.12.1 Design and operation comments

The biofilter is approximately 1 m in diameter and 2 m in height and contains two trays that hold the active medium. The trays can be rapidly changed for inspection and maintenance or for loading of medium with pre-adapted microbial population (Stewart and Thom, 1996). The system is designed so that most of the activity is located in the first layer of medium, while the second layer serves as a surge control.

The operation of the biofilter is almost entirely automatic, managed by a separate unit. Water is sprayed onto the trays, and nutrients can be added to the water at recommended intervals of one week. The only maintenance, besides loading and unloading rags in the drying chamber, is the weekly refill of the nutrient dispenser. The biofilter achieves good treatment success and stability through an unusually long empty bed retention time of 18 minutes and temperature control. Temperature control is possible for this system, in contrast to most, because of its small size and low flows.

9.12.2 Performance

Removal efficiency for the biofilter could not be evaluated with great accuracy because of the high variability of input concentrations and contaminant types (Stewart and Thom, 1996). However, input and output samples taken at intervals over a 6-week period showed a median removal of 99%. Operators report that the system significantly reduced VOC releases from their facility.

The system is being further applied to process and tank vents, to glycol dehydrators used in the oil industry, and to aerosol can puncturing and recycling operations. It has demonstrated that highly variable concentrations of contaminants can be treated if low flows allow very long detention times.

9.13 Soil biofilter for VOC removal in flexographic printing off-gases

This biofilter is located in an unused loading bay and parking area (Figure 9.14; Table 9.16). It treats off-gases from a flexographic printer. The biofilter was completed within 4 weeks from the time the South Coast Air Quality Management District (SCAQMD) permit was granted to avoid pollution penalties.

9.13.1 Design and operation comments

The biofilter treats approximately 10,200 m³ h⁻¹ (6000 cfm) of air containing 100 to 1000 ppmv of propanol and ethanol from three flexographic printing presses. The operation schedule is highly variable. One to three presses operate from 10 to 24 hours per day, 6 days per week. On average, the biofilter operates about 80 hours per week. Air flow is shut down during off time. The input air is humidified with fogger nozzles, and the bed has a sprinkler system. Both water inputs are actuated by the moisture sensors in the bed. The wall of the bed was made of concrete traffic dividers and required no building permit. The installed cost of the bed was $78,000.

The air permit required a minimum 70% control efficiency (capture efficiency × destruction efficiency). The regulatory agency recognized that biofilters create neither secondary air pollutants (NO_x or CO) nor extra greenhouse gases (CO_2 from thermal oxidizers). The printing presses are enclosed in rooms inside the building so that capture efficiency is greater

Figure 9.14 Picture of the biofilter showing the monitoring hoods on top of the biofilter. (Courtesy of Bohn Biofilter Corp.; Tucson, AZ.)

than 95%. The biofilter's removal efficiency should be 75% for an overall control efficiency of 70%.

Because monitoring emissions from open biofilter beds is difficult and can be very unreliable, this biofilter was equipped with a special sampling and capture device developed in cooperation with the South Coast Air Quality Management District. Ten hoods were located on top of the biofilter bed, i.e., where the air is leaving the surface. The hoods are $15 \times 90 \times 15$ cm high ($0.5 \times 3 \times 0.5$ ft) and placed at right angles to the air distribution pipes below to ensure that each hood crosses at least one distribution pipe. Each hood has a 5-mm hole far from the sampling port for pressure relief without mixing of atmospheric air. The hoods are connected by equal-length Teflon tubes to a manifold to ensure equal sampling from each location. The gas mixture flows from the manifold to a flame ionization detector (FID) for analysis. The hood locations are changed periodically. The x and y coordinates of the locations are calculated from a random number table to ensure that they are representative. Only randomized locations yield a true average. It should be noted that localized drying of the bed under the hoods can occur when significant moisture is provided by sprinklers located above the bed.

The SCAQMD also requires measurement of the air inlet temperature, bed back-pressure, and bed moisture content. The moisture potential is measured by three tensiometers placed across the bed. In contrast to organic

Table 9.16 Characteristics of the Flexographic Printing Biofilter

Owner and location	Sunshine Plastics; Montebello, CA
Builder	Bohn Biofilter Corp.; Tucson AZ
Type of air stream	Waste air from a flexographic printer
Year of installation	1997
Medium type and volume of medium	Soil, 486 m³; expected medium lifetime 30 years
Number of layers and height of medium	One layer of 0.9 m
Biofilter construction type	Open bed, above ground; maintained in concrete traffic divider, 540 m²
Humidification	Fogger nozzles in main air pipe; additional water sprinkler on the top of the bed; drainage is recirculated to the bed
Air flow rate	10,200 m³ h⁻¹
Empty bed residence time	2.5 min
Pressure drop	5 cm water gauge
Average bed temperature	18°C
Pollutants treated	Propanol, ethanol, acetone, varying concentrations 100 to 1000 ppm
Biofilter controls	Moisture monitored by three tensiometers placed in the bed which actuate fogger nozzles and sprinklers
Biofilter design and acceptance criterion	80% removal of organics monitored continuously
Approximate investment costs	$78,000
Approximate treatment cost per 1000 m³ off-gas treated	Electricity cost, primarily, i.e., $105 per year or $0.0026 per 1000 m³ off-gas treated (0.375 kW blower, 80 hours per week, 7 cts kWh⁻¹)
Performance	95% removal of total VOCs

media, soil is physically stable and its moisture properties are stable spatially with time. Hence, the moisture content measured with the tensiometers at three different points could be correlated to the overall moisture content of the bed and odor control performance.

9.13.2 Performance

The destruction/removal efficiency (DRE) of the biofilter is higher than 95% of the propanol-ethanol input. The widely varying input of VOC and the discontinuous operation of the biofilter have not caused any problems or changed the DRE appreciably. Water evaporation from the fogger nozzles cools the input air to 15°C year round so that summer temperatures have not affected the biofilter's performance (Bohn, 1997).

Figure 9.15   The Reemtsma biotrickling filter for odor control. (Courtesy of Zander Umwelt GmbH; Nuremberg, Germany.)

9.14 Biofilter or biotrickling filter?
Example of an intermittently watered
biotrickling filter for odor control

A tobacco company needed to reduce odors either by 90% or to a level lower than 100 OU m^{-3}. After evaluation of various pilot bioreactors, an intermittently watered biotrickling filter was installed (Figure 9.15; Table 9.17). In a sense, because of the intermittent trickling, this biotrickling filter is operated as a biofilter for most of the time.

9.14.1 Design and operation comments

Prior to installation of the full-scale reactor, pilot-scale tests were performed to ensure that the requirements of either 90% removal or odor reduction

Table 9.17 Characteristics of the Reemtsma Biotrickling Filter

Owner and location	Reemtsma; Berlin, Germany
Builder	Zander Umwelt GmbH; Nuremberg, Germany
Type of air stream	Cigarette production off-gas, odor treatment
Year of installation	1995
Medium type and volume of medium	Polyurethane foam, 500 m³
Number of layers and height of medium	One layer, 2.5 m high
Biofilter construction type	Six container units.
Humidification	No prehumidification
Air flow rate	160,000 m³ h⁻¹, downflow
Empty bed residence time	11 seconds
Pressure drop	400 Pa
Average bed temperature	40°C
Pollutants treated	Odors: 800–5500 OU m⁻³
Biofilter controls	Temperature, pressure drop, and water level controls
Biofilter design and acceptance criterion	90% odor removal or outlet air odor lower than 100 OU m⁻³
Approximate investment costs	4.3 million DM (1995, approximately $3.05 million) including ductwork, cooling towers, and heat exchangers
Approximate treatment cost per 1000 m³ off-gas treated	Operating costs of 160,000 DM per year, i.e., 0.114 DM per 1000 m³ off-gas treated (1997, $0.066 per 1000 m³ off-gas treated) (Loy, 1998)
Performance	>90% odor removal

below 100 OU m⁻³ could be met with the small (250 m²) footprint available on the roof of the facility. Another concern with the proposed technology was clogging of the bed by growing biomass, a problem commonly encountered in conventional biotrickling filters. The pilot tests showed that no clogging occurred within a reasonable time frame and that the odor removal was satisfactory at residence times as low as 10 seconds.

The bioreactor design includes a synthetic medium made of polyurethane foam. Compared to conventional biofiltration media, the polyurethane foam is very light and has a very high interfacial area: 20 kg m⁻³ and about 600 m² m⁻³, respectively. It has a reasonable water-holding capacity of 100 to 150 kg m⁻³, which allows operation with intermittent trickling of water (Loy et al., 1997). For use in the biotrickling filter, the foam is preshaped in cubes of 40 mm to facilitate installation into the reactor. To provide moisture, a small flow of water containing nutrients is recycled intermittently over the support. The pattern adopted in this case is sprinkling 5 to 15 minutes every hour. As is usual for Zander Umwelt GmbH biotrickling filters, the reactor is remotely controlled from an operator room using a modem.

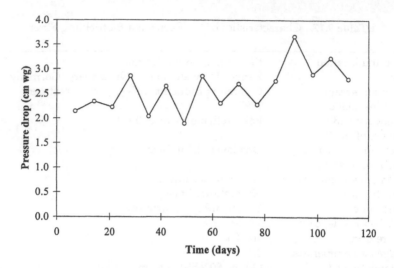

Figure 9.16   Over 100 days of pressure drop data for the Reemstma biotrickling filter. (Courtesy of Zander Umwelt GmbH; Nuremberg, Germany.)

9.14.2 Performance

Acclimation took place within 2 months, after which odor removal was continuously higher than 90%. Typical crude gas and clean gas data are 5400 and 400 OU m^{-3}, respectively, and removal consistently above 90%. No clogging of the filter bed occurred in the last 2 years, and pressure drop remained low and stable (Figure 9.16).

Appendix A

Symbols

Terms

A	m², ft²	biofilter area
a	m² m⁻³	interfacial area per volume unit
a_w	—	water activity
b	s⁻¹	biodegradation rate coefficient (first-order)
Bo	—	Bodenstein number
C_G	g m⁻³, ppm	concentration of contaminant (substrate) in air
C_{Gi}	g m⁻³, ppm	concentration of contaminant (substrate) in inlet air
C_{Go}	g m⁻³, ppm	concentration of contaminant (substrate) in outlet air
C_L	g m⁻³, Mol	concentration of contaminant (substrate) in water
C_L^*	g m⁻³, Mol	concentration of contaminant (substrate) in the water phase that is in equilibrium with the local concentration in the gas
C_{SW}	g m⁻³, Mol	concentration of contaminant (substrate) in solid-water phase
D	m² s⁻¹	diffusion coefficient
D_L	m² s⁻¹	axial dispersion coefficient
EBRT	s	empty bed residence time
EC	g m⁻³ h⁻¹	elimination capacity
EC_{max}	g m⁻³ h⁻¹	maximum elimination capacity
H	atm L mol⁻¹	Henry's Law coefficient
h	m	biofilter bed height
h_{Tot}	m	total biofilter bed height
K_0	g m⁻³ s⁻¹	zero-order reaction rate constant (Ottengraf, 1986 model)

251

K_1	s^{-1}	first order reaction rate constant (Ottengraf, 1986 model)
k	$g\ m^{-3}\ h^{-1}$, $Mol\ h^{-1}$	reaction rate, e.g., Michaelis-Menten
k_f	various	Freundlich adsorption isotherm constant
k_h	—	concentration partition coefficient between air and solids-water
K_H	—, $atm\ L\ mol^{-1}$	Henry's Law coefficient
K_i	$g\ m^{-3}$, Mol	inhibition constant in Haldane kinetics
K_M	$g\ m^{-3}$, Mol	half-saturation constant in Michaelis-Menten type of kinetics
K_{mass}	—	mass partition coefficient, between air and solids-water
k_{max}	$g\ m^{-3}\ h^{-1}$, $Mol\ h^{-1}$	maximum reaction rate in Michaelis-Menten type of kinetics
K_{ow}	—	octanol-water partition coefficient
K_S	$g\ m^{-3}$, Mol	half-saturation constant in Monod type of kinetics
k_t	h^{-1}	transfer rate coefficient for contaminant from air to water
M	kg	mass
m	—	dimensionless gas-liquid partition coefficient
n_{group}	—	occurrence of a particular group in QSAR models
Q	$m^3\ h^{-1}$, scfm	air flow rate
R	—	retardation factor; $v_i\ /v_{cont}$
R	$m^3\ bar\ kmol^{-1}\ K^{-1}$	universal gas constant, $R = 0.08314\ m^3\ bar\ kmol^{-1}\ K^{-1}$
RE	%	removal efficiency
r	m	radius of medium particle
t	s	time
T	K, °C	temperature
v	$m\ s^{-1}$	velocity
v_a	$m\ s^{-1}$	superficial or approach velocity
v_{cont}	$m\ s^{-1}$	average contaminant velocity
v_i	$m\ s^{-1}$	interstitial air velocity
V	m^3	volume
V_f	m^3	volume of filter bed
V_G	m^3	volume of air
V_L	m^3	volume of water
V_w	$m^3\ mol^{-1}$	partial molal volume of water
x	m	longitudinal distance within biofilter
X	$g\ m^{-3}$	concentration of biomass in the biofilm or in liquid phase
z	m	coordinate in the biofilm

Greek symbols

α, β, γ	various	constants in Johnson and Deshusses (1997) QSAR model
α_K^{group}, β_μ^{group}	various	Group contribution in Govind et al. (1997) QSAR model
δ	m	biofilm thickness
δ_{group}	—	group contribution in Johnson and Deshusses (1997) QSAR model
γ	m s^{-2}	surface tension of water
λ	m	effective biofilm thickness
μ	h^{-1}	microorganism-specific growth rate, e.g., in Monod kinetic
ψ	J kg^{-1}	water potential
τ	s	true detention time
θ	—	porosity, ratio of void volume to bed volume
χ	—	molecular connectivity index
$^1\chi^v$	—	first molecular connectivity index considering valence electrons

Superscripts, subscripts

ads	adsorbed
eff	effective
group	chemical group
i	inlet
i *or* j	pollutant species
max	maximum
O	oxygen
0	outlet
SW	solid-water phase
Tot	total

Abbreviations

ASCE	American Society of Civil Engineers
BTEX	benzene, toluene, ethyl benzene, xylene
CFC	chlorofluorohydrocarbons
cfm	cubic foot per minute
DMDS	dimethyl disulfide
DMS	dimethyl sulfide
EBRT	empty bed residence time
EC	elimination capacity
EPA	Environmental Protection Agency

GC/MS	gas chromatography with mass spectrometer detector
FAME	fatty acid methyl esters
FID	flame ionization detector
HAP	hazardous air pollutant
HCFC	hydrochlorofluorohydrocarbons
ID	internal diameter
MACT	maximum available control technology
MEK	methyl ethyl ketone
MIBK	methyl isobutyl ketone
NO_x	nitrogen oxides
PCE	perchloroethylene
PE	polyethylene
PLC	programmable logic controller
PLFA	phospholipid fatty acids
PM_{10}	particulate matter less than 10 microns
PP	polypropylene
ppmV	parts per million by volume
PVC	polyvinyl chloride
QSAR	quantitative structure activity relationship
RE	removal efficiency
scfm	standard cubic foot per minute
SVE	soil-vapor extraction
TCE	trichloroethylene
TPH	total petroleum hydrocarbon
VOC	volatile organic compound
WHO	World Health Organization

Appendix B

Selected elimination capacity values

Note that inlet concentrations, EBRT, and systems may not be directly comparable. These values are for information purposes only. The reader is directed to the references for further details. The critical loading reported here is the load at which the elimination capacity and the loading start to significantly differ. It corresponds approximately to 90 to 95% removal.

Elimination Values

Contaminant(s)	Biofilter medium	Critical load ($g\ m^{-3}\ h^{-1}$)	Maximum elimination capacity ($g\ m^{-3}\ h^{-1}$) (removal %)	Reference(s)
Acetone	Compost-based	229	229 (90%)	Briggs, 1996
Acetone	Compost-based	N/A	164	Ottengraf, 1987
Acetone	Compost-based	20–25	67	Johnson and Deshusses, 1997
Benzene	Compost-based	1	8	Johnson and Deshusses, 1997
Benzene	Compost-based + GAC	N/A	23	Eitner, 1990
BTEX	Carbon coated foam	N/A	41–55	De Filippi et al., 1993
BTEX	Sand	N/A	14–30	Kamarthi et al., 1994
BTEX	Carbon	N/A	15–44	Kamarthi et al., 1994
BTX	Compost/ perlite	35–40	50–60	Seed and Corsi, 1994
1,3-butadiene	Ceramic	N/A	30	van Groenestijn et al., 1995

Elimination Values

Contaminant(s)	Biofilter medium	Critical load (g m^{-3} h^{-1})	Maximum elimination capacity (g m^{-3} h^{-1}) (removal %)	Reference(s)
Butanol	Compost-based	30–40	70–80	Eitner, 1989
Butyl acetate	Compost-based	~10	40	Ottengraf et al., 1983
Butyl acetate	Compost-based	25–30	35	Johnson and Deshusses, 1997
Butyric acid	Compost-based	N/A	30	Eitner, 1989
Dimethyl disulfide	Compost/ pine mulch	10	10–12	Arpacioglu and Allen, 1996
Dimethyl sulfide	Wood bark	8–10	11–20	Smet et al., 1996
Dimethyl sulfide	Compost-based	N/A	70 (84%)	van Langenhove et al., 1996
Dioxan	Compost-based + GAC	N/A	11–13	Eitner, 1989
Ethanol	Granular carbon	N/A	156	Devinny and Hodge, 1995
Ethanol	Compost-based	80	150	Johnson and Deshusses, 1997
Ethanol/ 2-propanol	Compost-based	N/A	57	Ottengraf, 1987
Ethene	GAC	N/A	10	De Heyer et al., 1994
Ethyl acetate	Compost-based	8-10	25	Ottengraf and van den Oever, 1983
Ethyl acetate	Compost-based	180	200	Johnson and Deshusses, 1997
Ethylene	Ceramic	8–12	40–50	van Groenestijn et al., 1995
Gasoline vapors	Granular carbon	N/A	avg. 64, max. 119	Medina et al., 1995b
Hexane	Compost-based	1	5	Johnson and Deshusses, 1997
Hydrogen sulfide	Compost	100	130	Yang and Allen, 1994

Elimination Values

Contaminant(s)	Biofilter medium	Critical load $(g\ m^{-3}\ h^{-1})$	Maximum elimination capacity $(g\ m^{-3}\ h^{-1})$ (removal %)	Reference(s)
Isobutyl acetate	Compost-based	45	75	Johnson and Deshusses, 1997
Isopentane	Compost-based	2	8	Johnson and Deshusses, 1997
Jet fuel, JP-4	Peat	N/A	4–19	Ventera and Findlay, 1991
Jet fuel, JP-4	Sand	N/A	12	Ventera and Findlay, 1991
Jet fuel, JP-4	Compost	N/A	65	Ventera and Findlay, 1991
Methane	Glass rings	10–13	15–18	Sly et al., 1993
Methanol	Compost-based	42	N/A	Briggs, 1996
Methanol	Compost/perlite	10–20	301	Lee et al., 1996
Methanol	Compost-based	N/A	18	van Lith, 1989
Methanol	Compost-based	30–35	70	Johnson and Deshusses, 1997
Methanol	Compost/perlite	50–80	100–120	Shareefdeen et al., 1993
MEK	Compost-based	75–100	120	Deshusses, 1995
MIBK	Compost-based	15–18	25–30	Deshusses, 1995
MEK/MIBK	Compost-based	N/A	40/18	Deshusses, 1995
1-proapnol	Compost-based	120	150	Johnson and Deshusses, 1997
α-Pinene	Compost/perlite or compost/ GAC	N/A	35	Mohseni and Allen, 1996
Styrene	Perlite	62	62	Cox et al., 1997
Styrene	Peat	60–75	100	Togna and Folsom, 1992
Toluene	Peat	N/A	4–40	Auria et al., 1996
Toluene	Compost	N/A	100	Don and Feenstra, 1984
Toluene	Compost-based	8	15	Johnson and Deshusses, 1997

Elimination Values

Contaminant(s)	Biofilter medium	Critical load $(g\ m^{-3}\ h^{-1})$	Maximum elimination capacity $(g\ m^{-3}\ h^{-1})$ (removal %)	Reference(s)
Toluene	Compost-based	<10	20–25	Ottengraf and van den Oever, 1983
Toluene	Compost	30–40	45–55	Seed and Corsi, 1994
Xylene	Compost-based	10–15	25	Eitner, 1989

Appendix C

Conversion factors

Conversion Factors

To convert from	To	Multiply by
m	inch	39.37
m	foot	3.28
m	yard	1.094
mm	mil	39.4
m^2	square feet	10.76
m^3	liter	1000
m^3	gallon	264.2
m^3	cubic foot	35.31
m^3	cubic yard	1.308
$m^3 h^{-1}$	cubic foot per minute	0.5885
$m^3 h^{-1}$	gallons per minute	4.403
°C	degree Fahrenheit	(°C × 9/5) + 32
kg	pound	2.203
$g\ m^{-3}$	ppm_v	at 25°C: ($g\ m^{-3}$ × 24,776 / molecular weight of contaminant in $g\ mol^{-1}$)
ppm_v	$g\ m^{-3}$	at 25°C: (ppmv × molecular weight of contaminant in $g\ mol^{-1}$ /24,776)
kJ	BTU	0.9478
atm	cm wg	1033
Pa, $N\ m^{-2}$	cm wg	0.01020
cm wg	inch wg	0.3937
cm wg	inch of mercury	2.895×10^{-3}
Pa, $N\ m^{-2}$	pounds per square inch	1.450×10^{-4}
atm	pounds per square inch	14.70

Appendix D

Absolute humidity of air saturated with water

Absolute Humidity of Air Saturated With Water

Temperature (°C)	Mass of water per unit volume (g m⁻³)	Temperature (°C)	Mass of water per unit volume (g m⁻³)
0	4.85	32	33.8
5	6.80	34	37.6
10	9.41	36	41.7
12	10.7	38	46.2
14	12.1	40	51.1
16	13.6	42	56.5
18	15.4	44	62.5
20	17.3	46	68.8
22	19.4	48	75.7
24	21.8	50	83.2
26	24.4	55	104.6
28	27.2	60	130.5
30	30.4	70	198.4

Appendix E

Approximate conversion of selected currencies

Approximate conversion rates of selected currencies to U.S. dollar. The rate was recorded on June 15 of each year.

Currency Conversion

	1990	1991	1992	1993	1994	1995	1996	1997
1 British pound (£)	$1.71	$1.63	$1.86	$1.53	$1.52	$1.60	$1.54	$1.63
1 German mark (DM)	$0.59	$0.56	$0.64	$0.61	$0.61	$0.71	$0.65	$0.58
1 Dutch guilder (Dfl)	$0.52	$0.49	$0.57	$0.55	$0.54	$0.63	$0.58	$0.51
1 Swiss franc (Sfr)	$0.70	$0.65	$0.71	$0.69	$0.73	$0.86	$0.80	$0.69

Glossary

Absorption (compare to adsorption). Removal of a compound from a fluid because it passes into the volume of an adjacent phase.

Acclimation. Adjustment of a microbial cell or microbial culture to a sudden change in environmental conditions. In biofilters, operators seek rapid acclimation of the culture in a new biofilter to the presence of the substrate to be treated.

Adiabatic cooling. Cooling without transfer of heat to the surroundings. In biofiltration, this is often the cooling which occurs during humidification as a result of water evaporation.

Adiabatic. Process that occurs with no exchange of heat with its surroundings.

Adsorption (compare to absorption). Removal of a compound from a fluid because it sticks to the surface of a solid phase.

Adsorptive capacity. Amount of contaminant that can be adsorbed by an adsorbent material, expressed in mass of contaminant per unit weight or volume. The adsorptive capacity is dependent on the type of adsorbent, the type of contaminant, and the concentration of the contaminant.

Aerobic. Occurring in the presence of air; in this context, a biological process taking place in the presence of oxygen.

Agar plate. A petri dish filled with agar, a gel that is appropriate as a growth medium for microorganisms.

Alkalinity. Acid-neutralizing capacity of the medium. Alkalinity is defined with respect to a given endpoint pH, commonly pH 4.5. The alkalinity of the medium is the amount of acid which must be added in order to reduce the pH to the chosen endpoint value.

Anaerobic. Occurring in the absence of air; in this context, a biological process taking place in the absence of oxygen. Because of diffusion limitation, anaerobic conditions can exist in a biofilm, even if it is in contact with air containing 21% oxygen.

Autotroph. Organism that uses carbon dioxide as the sole source of carbon.

Bacteria. Single-celled microorganisms belonging to the prokaryotes. Bacteria are the dominant organisms in most biofilters.

Bioavailable. Being available to microorganisms for biodegradation. Compounds which are tightly adsorbed, held within an organic matrix, or are part of complexes may not be available for uptake by microorganisms.

Biofilm. A film, consisting of live microorganisms, materials such as polysaccharides exuded by microorganisms, and debris which forms on the surface of the medium in a biofilter.

Biomass. Biologically generated material, usually measured in a way which does not distinguish among types of material. Typical units are gram biomass per gram medium.

Bioscrubber. Device for removing contaminants from air in which the air contacts water in a spray chamber or trickling tower; the contaminant is dissolved, and the water is cleaned by biological treatment in a separate chamber.

Biotrickling filter. Device for removing contaminants from air in which the air is passed through an inert packing, which supports microorganisms, which degrade the pollutant, and through which there is a steady flow of recirculated water.

Bodenstein (Bo) number. Dimensionless number (vH/D_L); the product of the gas velocity and the height of the reactor divided by the axial dispersion coefficient. Reactors with Bodenstein numbers larger than 20 are considered to be plug flow reactors. Ideally mixed reactors have a Bo number approaching zero.

Buffer capacity. Degree to which a medium can resist pH changes. Precisely, the amount of base or acid which must be added per unit pH change in the medium.

Bulking agent. Material added to biofilter media for the purpose of increasing porosity, reducing compaction, and reducing the resistance to air flow. It generally consists of particles larger than those in the material to which it is added.

CFU (colony-forming units). Unit of measure used in a plate count for determining microbial density. It is generally presumed that a colony-forming unit is a single organism, so that the number of colonies represents the number of cells present in the sample, but the general term is often used in recognition of the fact that a colony may arise from a spore, or from a cluster of cells.

Cometabolism. Degradation of a compound by an organism which gains no energy or other advantage from the transformation. This may occur because of a fortuitous similarity between the structure of the cometabolized compound and a compound which the cell ordinarily uses (the growth substrate), so that the active site on an enzyme works for both compounds.

Connectivity index. One of several indices that can be calculated from the structure of a molecule, whose value depends on which atoms are connected and the manner in which they are connected.

Consortium. In this context, a group of microorganisms whose combined activities are responsible for degradation of a pollutant.

Constitutive enzyme. Enzyme that is produced at all times by the cell, regardless of the presence of its substrate.

Cosubstrate. In cometabolism, the compound that is degraded fortuitously and does not provide energy to the microorganisms; *see* Growth substrate.

Critical load. The biofilter load above which treatment efficiency falls substantially below 100%.

Cyst. Inactive resting stage in the life cycle of some microorganisms. Organisms can often survive for long periods under adverse conditions as cysts.

EBRT. *See* Empty bed residence time.

Elimination capacity. Amount of contaminant being removed in the biofilter, typically in units of grams per hour per cubic meter of biofilter volume.

Empty bed residence time (EBRT). Average residence time of air within the treatment bed, calculated as if the medium were not present. This is an overestimate of the true residence time, because the medium occupies some of the volume, but it is commonly used because the porosity of the medium is difficult to measure. (Compare to true residence time.)

Eukaryote. Cell that contains nucleus surrounded by a membrane. Eukaryotes are generally more complex than prokaryotes. The group includes protozoa which are bacterial predators.

Fatty acid methyl ester analysis (FAME). Analytical technique in which fatty acids are extracted from a sample of cell culture and analyzed by gas chromatography. The types of fatty acids found may provide clues to the types of cells present and the conditions under which they are growing.

FID. *See* Flame ionization detector.

Finite differences approximation. Method used for calculation of approximate solutions to differential equations, in which the domain of the equation is divided into very small increments, and linear approximation is used within each increment.

Flame ionization detector (FID). Detector which can measure the amount of a compound in a stream of gas, often used as the detector of a gas chromatograph. The gas sample is passed through a flame, and the ions created are detected.

Forced draft. Design in which air is forced under pressure through a biofilter by a blower upstream; *see* Induced draft.

Fungus. Heterotrophic eukaryote with a cell wall.

Gas chromatograph. Instrument used to measure concentrations of compounds and, indirectly, to identify them. Vapors are passed through an adsorbent porous medium, which causes the compounds to separate into individual "peaks", the shape of which is measured by a detector. The size of the peak indicates total concentration, and testing of standards can identify the peaks.

Gram test. Staining technique for classification of microorganisms as gram positive or gram negative based on the retention of crystal violet dye after decolorization of the bacteria with alcohol.

Growth substrate. In cometabolism, the compound that provide energy for the growth of the microorganisms; *see* Cosubstrate.

Heterotroph. Organism obtaining energy and carbon from organic compounds.

Hydrophilic. Having the characteristic of adsorbing water or dissolving in water. Polar minerals are hydrophilic, and water will spread on their surfaces rather than beading.

Hydrophobic. Having the characteristic of repelling or not adsorbing water, or having low solubility in water. Non-polar organic materials are hydrophobic, and water beads on such surfaces rather than spreading.

Hyphae. Filamentous extensions grown by microorganisms, most commonly fungi.

Induced draft. Design in which air is pulled by vacuum through a biofilter by a blower downstream; *see* Forced draft.

Inoculum. Culture of microorganisms added to a system to provide cells of desirable species. Biofilters are sometimes inoculated with pure cultures prepared in the laboratory and sometimes with cultures prepared from soil or sewage sludge.

Interstitial velocity. Air flow velocity measured for particles of air within the pores of the medium.

Lag phase. Initial phase in a bacterial culture where no population growth is observed. The cells may be metabolically active but are not increasing in number.

Laminar layer. Layer of air or water in which flow is smooth and non-turbulent, usually a thin region near the surface of a solid.

Lithotroph. Organism that can obtain energy from the oxidation of inorganic compounds such as hydrogen sulfide.

Load cell. Device that measures the weight of the load it carries. Load cells are mounted beneath biofilters to monitor their total weight in order to follow changes in water content.

Macrokinetic. Referring to overall kinetics of pollutant elimination for a bioreactor or part of it. Includes the effects of mass transfer and other phenomena.

Mass loading rate. Amount of contaminant entering the biofilter, typically in units of grams per hour per cubic meter of biofilter volume.

Medium (plural, media). In this text, the material used to support the microorganisms in the biofilter. Also referred to as packing material.

Mesophile. Organism whose optimum growth temperature is between 20 and 40°C.

Mesophilic. Well-adapted to growth and activity in environments of moderate temperature; used to describe microorganisms.

Metabolism. Sum of all processes performed by a living organism.

Michaelis-Menten equation. Relationship describing the rate of transformation of a substrate by an enzyme or microorganism, under the assumption that the amount of biomass present is constant.

Microkinetic. Referring to kinetics of pollutant biodegradation at the microscopic scale, generally be described by models such as Michaelis-Menten or Monod or other more complex relationships.

Mineralization. Conversion of organic pollutant to inorganic molecules, e.g., carbon dioxide, water, and sulfate.

Moisture content. Amount of water in a material. This may be defined as percent of wet weight, ranging from 0% for perfectly dry soil or compost medium to 100% for pure water. However, it may also be defined as percent of dry weight and range from 0% for perfectly dry soil to values greater than 100% in very wet soils. Care should be taken to determine the convention being used in publications and to define the convention used when reporting data.

Molecular connectivity index. Index calculated from molecular structures which depends on the characteristics of atoms "connected" or bonded to each other. Indices may depend on atoms that are adjacent or are connected in groups of three or more, and are primarily used to develop quantitative structure activity relationships.

Monod equation. Relationship describing how the rate of biomass growth depends on the concentration of substrate available.

Odor unit (OU m^{-3}). A measure of the intensity of odor. The dilution ratio necessary to reduce the concentration to the point at which the odor is no longer detectable is the number of odor units present. The units of m^{-3} are commonly included, although they are not strictly justified. Equivalently, the concentration of the odor-causing compound measured in multiples of the odor-detection threshold concentration.

Partition coefficient. Ratio of concentrations for a compound in two adjacent phases at equilibrium. Biofilter designers are often concerned with the partition coefficient between the air and the water, also called the Henry's coefficient, which is the ratio of the air concentration to the water concentration for a contaminant at equilibrium.

Perlite. Highly porous clay mineral sometimes used as a biofilter medium.

Phospholipid fatty acid analysis (PLFA). Analytical technique in which phospholipids are extracted from a sample of cell culture, and analyzed by gas chromatography. The types of fatty acids found may provide clues to the types of cells present and the conditions under which they are growing.

PLC. *See* Programmable logic controller.

Plenum. Region that receives flows of air from several sources or distributes a flow to several sources.

PLFA. *See* Phospholipid fatty acid analysis.

Polysaccharide gel. Mucus-like material exuded by microorganisms, and the major component of the biofilm.

ppb. Parts per billion; in aqueous samples, ppb corresponds to μg L^{-1}.

ppm. Parts per million; in aqueous samples, ppm corresponds to mg L^{-1}.

ppmv. Parts per million on a volume basis. Used extensively for gaseous samples. The conversion from ppmv to a mass concentration (g m^{-3}) requires the knowledge of the molecular weight of the compound and the use of the ideal gas law (see Appendix C).

Pressure-compensated irrigation hose. Water irrigation device placed within or on top of a biofilter bed to maintain optimal moisture content. Small holes are spaced equally along the length of the hose where individual pressure compensated valves regulate water flow into the filter bed. *See* Soaker hose.

Programmable logic controller (PLC). A computer-controlled system which can be used to collect data automatically and to control operations in a biofilter.

Prokaryote. Cell whose genetic material is not surrounded by a nuclear membrane.

Psychrophiles. Organisms which are well adapted for growth at low temperatures.

Psychrophilic. Well-adapted to growth and activity in cold environments; used to describe microorganisms.

Quantitative structure activity relationship (QSAR). A relationship, typically determined by regression, between the structural characteristics of a molecule and various chemical or biological parameters, such as vapor pressure, solubility, adsorption coefficient, or biodegradability.

Regenerative blower. A blower with a specially designed impeller to provide higher pressures or higher vacuums than conventional blowers

Removal efficiency. Fraction of contaminant entering the biofilter which is removed from the air stream, typically expressed as percent.

Retardation factor. Ratio between the rate of movement of the air in a biofilter and the rate of movement of the contaminant. If the contaminant is strongly adsorbed, the retardation factor will be high.

Soaker hose. Type of water irrigation device placed within or on top of a biofilter bed to maintain optimal moisture content. Small holes are spaced equally along the length of the hose to allow water to drip into the filter bed.

Soil vapor extraction. Technique for cleaning polluted soils in which air is blown through the soil from wells in order to evaporate contaminants.

Stoichiometry. Study of the molar ratios in which compounds are consumed or created in chemical reactions.

Surface loading rate. Volume of air passing through the input surface of the bed, typically in units of cubic meters of air per second per square meter of bed surface. The combined units are meters per second (m s^{-1}), and the surface loading rate can be interpreted as the average velocity at which the air is approaching the bed.

SVE. *See* Soil vapor extraction.

Thermophile. Organisms that grow best at temperatures above 40°C.

Thermophilic. Well-adapted to growth and activity in hot environments; used to describe microorganisms.

Total petroleum hydrocarbons (TPH). Commonly measured as the general response of a flame ionization detector.

Tracer. Inert substance used to determine residence time distribution and air flow pattern. Methane and smoke are often used as tracers.

True residence time. Average residence time of air within the treatment bed.

Volatile organic compound (VOC). An organic compound with a high vapor pressure which correspondingly may be present in the air in high concentrations.

Volumetric loading rate. Volume of air passing through the biofilter, typically in units of cubic meters of air per second per cubic meter of bed volume.

Wall effects. Phenomena occurring at the walls of a biofilter, usually short-circuiting of the air flow. Wall effects are of particular concern in bench-scale testing, where the higher ratio of perimeter to cross-sectional area means the effects can be large relative to overall biofilter activity.

Water gauge (wg). Scale for measuring pressure, determined as the height of a column of water which will exert that pressure.

References

Acuna, E., Auria, R., Pineda, J., Perez, F., Morales, M., and Revah, S., Studies on the microbiology and kinetics of a biofilter used to control toluene emissions, in *Proceedings of the 89th Annual Meeting and Exhibition of the Air and Waste Management Association*, Air and Waste Management Association, Pittsburgh, PA, 1996.

Alexander, M., *Introduction to Soil Microbiology*, Second ed., John Wiley & Sons, New York, 1977.

Alexander, M., *Biodegradation and Bioremediation*, Academic Press, New York, 1994.

Allen, E.R. and Yang, Y., Biofiltration control of hydrogen sulfide emissions, in *Proceedings of the 84th Annual Meeting and Exhibition of the Air and Waste Management Association*, Air and Waste Management Association, Pittsburgh, PA, 1991.

Allen, E.R. and Phatak, S., Control of organo-sulfur compound emissions using biofiltration-methyl mercaptan, in *Proceedings of the 86th Annual Meeting and Exhibition of the Air and Waste Management Association*, Air and Waste Management Association, Pittsburgh, PA, 1993.

Allen, P.J. and van Til, T.S., Installation of a full scale biofilter for odor reduction at a hardboard mill, in *Proceedings of the 1995 Conference on Biofiltration (an Air Pollution Control Technology)*, Hodge, D.S. and Reynolds, F.E., Eds., The Reynolds Group, Tustin, CA, 1995, p. 31.

Alonso, C., Suidan, M.T., Sorial, G.A., Smith, F.L., Biswas, P., Smith, P.J., and Brenner, R.C., Gas treatment in trickle-bed biofilters: biomass, how much is enough?, *Biotechnol. Bioeng.*, 54(6), 583, 1997.

Amann, R.I., Ludwig, W., and Schleifer, K.H., Phylogenetic identification and *in situ* detection of individual microbial cells without cultivation, *Microbiol. Rev.*, 59(1), 143, 1995.

American Society of Civil Engineers, *J. Environ. Eng.*, 123(6), 1997.

Anzion, C.J.M., Smith, W., and Valk, C.J., Management and monitoring of biofilters in composting plants, in *Biological Waste Gas Cleaning, Proceedings of an International Symposium*, Prins, W.L. and van Ham, J., Eds., VDI Verlag GmbH, Duesseldorf, Germany, 1997, p. 289.

Apel, W.A., Barnes, J.M., and Barrett, K.B., Biofiltration of nitrogen oxides from fuel combustion gas streams, in *Proceedings of the 88th Annual Meeting of the Air and Waste Management Association*, Air and Waste Management Association, Pittsburgh, PA, 1995.

APHA, *Standard Methods for the Examination of Water and Wastewater*, 19th ed., American Public Health Association, New York, 1995.

273

Arpacioglu, B.C. and Allen, E.R., Control of organo-sulfur compound emissions using biofiltration: dimethyl disulfide, in *Proceedings of the 89th Annual Meeting and Exhibition of the Air and Waste Management Association*, Air and Waste Management Association, Pittsburgh, PA, 1996.

Atlas, R.M. and Bartha, R., *Microbial Ecology — Fundamentals and Applications*, Third ed., The Benjamin/Cummings Publishing Company, New York, 1993.

Auria, R.M., Morales, M., Acuna, E., Perez, F., and Reva, S., Biofiltration of toluene vapors: start up and gaseous ammonia addition, in *Proceedings of the 1996 Conference on Biofiltration (an Air Pollution Control Technology)*, Reynolds, F.E., Ed., The Reynolds Group, Tustin, CA, 1996, p. 134.

Austin, D.K., Sandosham, A., Ventola, R.J., and Holmes, J., Four case studies: technical and economic feasibility of biofiltration as a technology for VOC and odor control, in *Proceedings of the 1995 Conference on Biofiltration (an Air Pollution Control Technology)*, Hodge, D.S. and Reynolds, F.E., Eds., The Reynolds Group, Tustin, CA, 1995, p. 295.

Baltzis, B.C. and Androutsopoulou, H., A study on the response of biofilters to shock-loading, in *Proceedings of the 87th Annual Meeting and Exhibition of the Air and Waste Management Association*, Air and Waste Management Association, Pittsburgh, PA, 1994.

Barnes, J.M., Apel, W.A., and Barrett, K.B., Removal of nitrogen oxides from gas streams using biofiltration, *J. Hazardous Mater.*, 41(2–3), 315, 1995.

Berndt, M. and Mildenberger, H.J., Biofiltersysteme zur Geruchsbeseitigung und zur Reduzierung von Organika-Emissionen — Betriebserfahrungen, Sanierungs — und Optimierungsmaßnahmen, in *Biologische Abgasreinigung*, VDI Berichte 1104, VDI Verlag GmbH, Duesseldorf, Germany, 1994, p. 373.

Bishop, D.F. and Govind, R., Development of novel biofilters for treatment of volatile organic compounds, in *Biological Unit Processes for Hazardous Waste Treatment*, *Proceedings of the Third International In Situ and On-Site Bioreclamation Symposium*, Hinchee, R.E., Skeen, R.S., and Sayles, G.D., Eds., Battelle Press, Columbus, OH, 1995, p. 219.

Bligh, E.G. and Dyer, W.J., A rapid method of total lipid extraction and purification, *Can. J. Biochem. Physiol.*, 8, 911, 1959.

Bodker, J. and Rydin, S., Biological waste gas treatment using a vertical filter, in *Proceedings of the 1996 Conference on Biofiltration (an Air Pollution Control Technology)*, Reynolds, F.E., Ed., The Reynolds Group, Tustin, CA, 1996, p. 39.

Boethling, R.S., Howard, P.H., Meylan, W., Stiteler, W., Beauman, J., and Tirado, N., Group contribution method for predicting probability and rate of aerobic biodegradation, *Environ. Sci. Technol.*, 28, 459, 1994.

Bohn, H.L., Soil and compost filters of malodorant gases, *J. Air Pollut. Control Assoc.*, 25, 953, 1975.

Bohn, H.L., Biofilter media, in *Proceedings of the 89th Annual Meeting and Exhibition of the Air and Waste Management Association*, Air and Waste Management Association, Pittsburgh, PA, 1996.

Bohn, H.L., Personal communication, Bohn Biofilter Corp., Tucson, AZ, 1997.

Bohn, H.L. and Bohn, R.K., Soil beds weed out air pollutants, *Chem. Eng.*, 95(6), 73, 1988.

Boyette, R.A., Williams, T., and Wynne, D., Biofiltration demonstration for controlling volatile organic emissions, in *Proceedings of the 1995 Conference on Biofiltration (an Air Pollution Control Technology)*, Hodge, D.S. and Reynolds, F.E., Eds., The Reynolds Group, Tustin, CA, 1995, p. 325.

Braun-Lulleman, A., Johannes, C., Majcherczyk, A., and Hutterman, A., The use of white-rot fungi as active biofilters, in *Biological Unit Processes for Hazardous Waste Treatment, Proceedings of the Third International In Situ and On-Site Bioreclamation Symposium*, Hinchee, R.E., Skeen, R.S., and Sayles, G.D., Eds., Battelle Press, Columbus, OH, 1995, p. 235.

Briggs, T.G., Biofiltration of air containing volatile organic chemicals from batch manufacturing operations, in *Proceedings of the 1996 Conference on Biofiltration (an Air Pollution Control Technology)*, Reynolds, F.E., Ed., The Reynolds Group, Tustin, CA, 1996, p. 69.

Brock, T.D. and Madigan, M.T., *Biology of Microorganisms*, Sixth ed., Prentice Hall, Englewood Cliffs, NJ, 1991.

Brook, R.R., Stiver, W.H., and Zytner, R.G., Effect of nitrogen sources on the biodegradation of diesel fuel in unsaturated soil, in *Proceedings of the 1997 CSCE-ASCE Environmental Engineering Conference*, American Society of Civil Engineers, New York, 1997, p. 975.

BUWAL, *Ablufreinigung mit Biofiltern and Biowaschern*, BUWAL Schriftenreihe Umwelt Nr.204, BUWAL Editor, Bern, Switzerland, 1993.

Carlson, D.A. and Leiser, C.P., Soil beds for the control of sewage odors, *J. Water Pollut. Control Fed.*, 38(5), 829, 1966.

Chang, A.N. and Yoon, H., Biofiltration of gasoline vapors, in *Proceedings of the 1995 Conference on Biofiltration (an Air Pollution Control Technology)*, Hodge, D.S. and Reynolds, F.E., Eds., The Reynolds Group, Tustin, CA, 1995, p. 123.

Chang, A.N. and Devinny, J.S., Biofiltration of JP4 jet fuel vapors, in *Proceedings of the 1996 Conference on Biofiltration (an Air Pollution Control Technology)*, Reynolds, F.E., Ed., The Reynolds Group, Tustin, CA, 1996, p. 142.

Characklis, W.G. and Marshall, K.C., *Biofilms*, John Wiley & Sons, New York, 1990.

Cherry, R.S. and Thompson, D.N., The shift from growth to nutrient-limited maintenance kinetics during acclimation of a biofilter, *Biotechnol. Bioeng.*, 56(3), 330, 1997.

Choi, D.S., Webster, T.S., Chang, A.N., and Devinny, J.S., Quantitative structure-activity relationships for biofiltration of volatile organic coupounds, in *Proceedings of the 1996 Conference on Biofiltration (an Air Pollution Control Technology)*, Reynolds, F.E., Ed., The Reynolds Group, Tustin, CA, 1996, p. 231.

Christensen, B.E. and Characklis, W.G., Physical properties of biofilms, in *Biofilms*, Characklis, W.G. and Marshall, K.C., Eds., John Wiley & Sons, New York, 1990, p. 93.

Corsi, R.L. and Seed, L.P., Biofiltration of BTEX-contaminated gas streams: laboratory studies, in *Proceedings of the 87th Annual Meeting and Exhibition of the Air and Waste Management Association*, Air and Waste Management Association, Pittsburgh, PA, 1994.

Cox, H.H.J. and Deshusses, M.A., Increasing the stability of biotrickling filters by using protozoa, in *Biological Waste Gas Cleaning, Proceedings of an International Symposium*, Prins, W.L. and van Ham, J., Eds., VDI Verlag GmbH, Duesseldorf, Germany, 1997a, p. 233.

Cox, H.H.J., and Deshusses, M.A., Evaluation of different technologies to control biomass growth in biotrickling filters for waste air treatment, presented at the Emerging Technologies in Hazardous Waste Managment IX, Special Symposium, Industrial and Engineering Chemistry Division, American Chemical Society, Pittsburgh, PA, 1997b.

Cox, H.H.J., Houtman, J.H.M., Doddema, H.J., and Harder, W., Enrichment of fungi and degradation of styrene in biofilters, *Biotechnol. Lett.*, 15(7), 737, 1993.

Cox, H.H.J., Styrene Removal from Waste Gas by the Fungus *Exophiala jeanselmei* in a Biofilter, Ph.D. thesis, University of Groningen, The Netherlands, 1995.

Cox, H.H.J., Moerman, R.E., van Baalen, S., van Heiningen, W.N.M., Doddema, H.J., and Harder W., Performance of a styrene-degrading biofilter containing the yeast *Exophiala jeanselmei*, *Biotechnol. Bioeng.*, 53(3), 259, 1997.

Cussler, E.L., *Diffusion: Mass Transfer in Fluid Systems*, 2nd ed., Cambridge University Press, New York, 1997.

Davidova, Y.B., Schroeder, E.D., and Chang, D.P.Y., Biofiltration of nitric oxide, in *Proceedings of the Air and Waste Management Association's 90th Annual Meeting and Exhibition*, Air and Waste Management Association, Pittsburgh, PA, 1997.

de Beer, D., Stoodley, P., Roe, F., and Lewandowski, Z., Effects of biofilm structures on oxygen distribution and mass transport, *Biotechnol. Bioeng.*, 43(11), 1131, 1994.

de Castro, A., Allen, D.G.G., and Fulthorpe, R.R., Characterization of the microbial population during biofiltration and the influence of the inoculum source, in *Proceedings of the 1996 Conference on Biofiltration (an Air Pollution Control Technology)*, Reynolds, F.E., Ed., The Reynolds Group, Tustin, CA, 1996, p. 164.

de Castro, A., Allen, D.G.G., and Fulthorpe, R.R., Characterization of the microbial population during biofiltration and the influence of the inoculum source, in *Proceedings of the Air and Waste Management Association's 90th Annual Meeting and Exhibition*, Air and Waste Management Association, Pittsburgh, PA, 1997.

de Felippi, L.J., Koch, M.B., Voellinger, C.M., Winstead, D., and Lupton, F.S., A biological air treatment system based upon the use of structured carbon biomass support, presented at the IGT Symposium on Gas, Oil, and Environmental Biotechnology, Colorado Springs, CO, 1993.

de Heyer, B., Overmeire, A., van Langenhove, H., and Verstraete, W., Biological treatment of waste gas containing the poorly water-soluble compound ethene using a dry granular activated carbon biobed, in *Biologische Abgasreinigung*, VDI Berichte 1104, VDI Verlag GmbH, Duesseldorf, Germany, 1994, p. 301.

Deshusses, M.A., Biodegradation of Mixtures of Ketone Vapours in Biofilters for the Treatment of Waste Air, Ph.D. thesis, Swiss Federal Institute of Technology, Zurich, 1994.

Deshusses, M.A., Transient behavior of biofilters: start-up, carbon balances, and interactions between pollutants, *J. Environ. Eng.*, 123(6), 563, 1997a.

Deshusses, M.A., Biological waste air treatment in biofilters, *Curr. Opin. Biotechnol.*, 8(3), 335, 1997b.

Deshusses, M.A. and Dunn, I.J., Modeling experiments on the kinetics of mixed solvent removal from waste gas in a biofilter, in *Proceedings of the 6th European Congress in Biotechnology*, Alberghina, L., Frontali, L., and Sensi, P., Eds., Elsevier, The Netherlands, 1994, p. 1191.

Deshusses, M.A., Hamer, G., and Dunn, I.J., Behavior of biofilters for waste air biotreatment. 1. Dynamic model development, *Environ. Sci. Technol.*, 29(4), 1048, 1995a.

Deshusses, M.A., Hamer, G., and Dunn, I.J., Behavior of biofilters for waste air biotreatment. 2. Experimental evaluation of a dynamic model, *Environ. Sci. Technol.*, 29(4), 1059, 1995b.

Deshusses, M.A., Hamer, G., and Dunn, I.J., Transient-state behavior of a biofilter removing mixtures of vapors of MEK and MIBK from air, *Biotechnol. Bioeng.*, 49(5), 587, 1996.

Deshusses, M.A., Johnson, C.T., Hohenstein, G.A., and Leson, G., Treating high loads of ethyl acetate and toluene in a biofilter, in *Proceedings of the 90th Annual Meeting and Exhibition of the Air and Waste Management Association*, Air and Waste Management Association, Pittsburgh, PA, 1997.

Devinny, J.S., Soil water content and air stripping, in *Environmental Engineering, Proceedings of the 1989 Specialty Conference*, American Society of Civil Engineers, New York, 1989, p. 555.

Devinny, J.S. and Hodge, D.S., Formation of acidic and toxic intermediates in over-loaded ethanol biofilters, *J. Air Waste Manage. Assoc.*, 45(2), 125, 1995.

Devinny, J.S., Medina, V.F., and Hodge, D.S., Bench testing of fuel vapor treatment by biofiltration, presented at National Research and Development Conference on the Control of Hazardous Materials, Hazardous Materials Control Research Institute, Anaheim, CA, February 20–22, 1991.

Devinny, J.S., Webster, T.S., Torres, E.M., and Basrai, S.S., Biofiltration for removal of PCE and TCE vapors from contaminated air, *Hazardous Waste Hazardous Mater.*, 12(3), 283, 1995.

Devinny, J.S., Schwarz, B.C.E., Chitwood, D., and Tsotsis, T.T., Biofiltration of per-chloroethylene, presented at Emerging Technologies in Hazardous Waste Managment IX, Special Symposium, Industrial and Engineering Chemistry Division, American Chemical Society, Pittsburgh, PA, 1997a.

Devinny, J.S., Choi, D.S., and Webster, T.S., Quantitative structure-activity relation-ships for prediction of removal efficiency in air-phase biofilters, presented at Emerging Solutions to VOC and Air Toxics Control, Air and Waste Management Association, San Diego, CA, February 26–28, 1997b.

Dharmavaram, S., Biofiltration, a lean emissions abatement technology, in *Proceedings of the 84th Annual Meeting and Exhibition of the Air and Waste Management Association*, Air and Waste Management Association, Pittsburgh, PA, 1991.

Diks, R.M.M., Ottengraf, S.P.P., and van den Oever, A.H.C., The influence of NaCl on the degradation rate of dichloromethane by *Hyphomicrobium* sp., *Biodegradation*, 5, 129, 1994.

Dombroski, E.C., Gaudet, I.D., and Coleman, R.N., Reduction of odorous and toxic emissions from Kraft pulp mills using biofilters, in *Proceedings of the 1995 Conference on Biofiltration (an Air Pollution Control Technology)*, Hodge, D.S. and Reynolds, F.E., Eds., The Reynolds Group, Tustin, CA, 1995, p. 287.

Don, J.A. and Feestra, L., Odor abatement through biofiltration, in *Proceedings of Characterization and Control of Odouriferous Pollutants in Process Industries*, Le Neuve, Belgium, 1984.

Dragan, J., *The Soil Chemistry of Hazardous Materials*, Hazardous Material Control Research Institute, Silver Springs, MD, 1988.

Dragt, A.J, Opening Address, in *Biotechniques for Air Pollution Abatement and Odour Control Policies*, Dragt A.J. and van Ham, J., Eds., Elsevier Science, Amsterdam, 1992, p. 3.

Durham, D.R., Marshall, L.C., Miller, J.G., and Chmurny, A.B., Characterization of inorganic biocarriers that moderate system upsets during fixed-film biotreatment processes, *Appl. Environ. Microbiol.*, 60(9), 3329, 1994.

Eitner, D., Vergleich von Biofiltermedien anhand mikrobiologischer und boden-physikalisher Kenndaten, in *Biologische Abgasreinigung*, VDI Berichte 735, VDI Verlag GmbH, Duesseldorf, Germany, 1989, p. 191.

Eitner, D., Biofilter in der praxis, in *Biologische Abluftreinigung*, Expert Verlag Ehningen bei Boeblingen, Germany, 1990, p. 55.

Enzien, M.V., Picardal, F., Hazen, T.C., Arnold, R.G., and Fliermans, C.B., Reductive dechlorination of trichloroethylene and tetrachloroethylene under aerobic conditions in a sediment column, *Appl. Environ. Microbiol.*, 60(6), 2200, 1994.

Ergas, S.J., Schroeder, E.D., and Chang, D.P.Y., Control of dichloromethane, trichloroethene, and toluene by biofiltration, in *Proceedings of the 86th Annual Meeting and Exhibition of the Air and Waste Management Association*, Air and Waste Management Association, Pittsburgh, PA, 1993.

Ergas, S.J., Schroeder, E.D., Chang, D.P.Y., and Scow, K., Spatial distribution of microbial populations in biofilters, in *Proceedings of the 87th Annual Meeting and Exhibition of the Air and Waste Management Association*, Air and Waste Management Association, Pittsburgh, PA, 1994.

Ergas, S.J., Schroeder, E.D., Chang, D.P.Y., and Morton, R., Control of VOC emissions from a POTW using a compost biofilter, *Water Environ. Res.*, 67(7), 816, 1995.

ERI, (P. Petro and K. Romstad), personal communication, Environmental Resolutions, Inc., Novato, CA, 1998.

Eweis, J.B., Chang, D.P.Y., Schroeder, E.D., Scow, K.M., Morton, R.L., and Caballero, R., Meeting the challenge of MTBE biodegradation, in *Proceedings of the 90th Annual Meeting and Exhibition of the Air and Waste Management Association*, Air and Waste Management Association, Pittsburgh, PA, 1997.

Farmer, R.W., Chen, J.S., Kopchynski, D.M., and Maier, W.J., Reactor switching: proposed biomass control strategy for the biofiltration process, in *Biological Unit Processes for Hazardous Waste Treatment, Proceedings of the Third International In Situ and On-Site Bioreclamation Symposium*, Hinchee, R.E., Skeen, R.S., and Sayles, G.D., Eds., Battelle Press, Columbus, OH, 1995, p. 243.

Finn, L. and Spencer, R., Managing biofilters for consistent odor and VOC treatment, *BioCycle*, 38(1), 40, 1997.

Fouhy, K., Cleaning waste gas naturally, *Chem. Eng.*, 99(12), 41, 1992.

Fucich, W.J., Yang, Y., and Togna, A.P., Biofiltration for control of carbon disulfide and hydrogen sulfide vapors, in *Proceedings of the 90th Annual Meeting and Exhibition of the Air and Waste Management Association*, Air and Waste Management Association, Pittsburgh, PA, 1997.

Fuller, R., Web site — http://www.wastewater.net, 1997.

Fuller, R., Personal communication, U.S. Filter, Naperville, IL, 1998.

Furusawa, N., Togashi, I., Hirai, M., Shoda, M., and Kubota, H., Removal of hydrogen sulfide by a biofilter with fibrous peat, *J. Ferment. Technol.*, 62(6), 589, 1984.

Gerrard, A.M., Economic design of biofilter systems, *J. Chem. Technol. Biotechnol.*, 68(4), 377, 1997.

Geyer, J.R., Mabury, S.A., and Crosby, D.G., Rice field surface microlayers: collection, composition and pesticide enrichment, *Environ. Toxicol. Chem.*, 15(10), 1676, 1996.

Gibbons, M.J. and Loehr, R.C., Effect of media nitrogen concentration on biofilter performance, *J. Air Waste Manage. Assoc.*, 48(3), 475, 1998.

Gilmore, G.L. and Briggs, T.G., Experiences with a compost biofilter for VOC control from batch chemical manufacturing operations, in *Proceedings of the 90th Annual Meeting and Exhibition of the Air and Waste Management Association*, Air and Waste Management Association, Pittsburgh, PA, 1997.

Govind, R., Lai, L., and Dobbs, R., Integrated model for predicting the fate of organics in wastewater treatment plants, *Environ. Prog.*, 10(1), 13, 1991.

Govind, R., Desai, S., and Bishop, D.F., Control of biomass growth in biofilters, in *Proceedings of the Fourth International In Situ and On-Site Bioremediation Symposium*, Vol. 5, Battelle Press, Columbus, OH, 1997a, p. 195.

Govind, R., Wang, Z., and Bishop, D.F., Biofiltration kinetics for volatile organic compounds and the development of a structure-biodegradability relationship, in *Proceedings of the 90th Annual Meeting and Exhibition of the Air and Waste Management Association*, Air and Waste Management Association, Pittsburgh, PA, 1997b.

Graham, J.R., GAC based gas phase biofiltration, in *Proceedings of the 1996 Conference on Biofiltration (an Air Pollution Control Technology)*, Reynolds, F.E., Ed., The Reynolds Group, Tustin, CA, 1996, p. 85.

Guckert, J.B. and White, D., Phospholipid, ester-linked fatty acid analysis in microbial ecology, in *Proceedings of the Fourth International Symposium on Microbial Ecology*, Ljubljara, August 24–29, 1986.

Guckert, J.B., Hood, M.A., and White, D.C., Phospholipid ester-linked fatty acid profile changes during nutrient deprivation of *Vibrio chloerae*: increases in the trans/cis ratio and proportions of cyclopropyl fatty acids, *Appl. Environ. Microbiol.*, 52, 794, 1986.

Harremoes, P., Half-order reactions in biofilm and filter kinetics, *VATTEN*, 33(2), 122, 1977.

Harris, R.F. and Sommers, L.E., Plate-dilution frequency technique for assay of microbial ecology, *Appl. Microbiol.*, 16, 330, 1968.

Hodge, D.S., Personal communication, The Reynolds Group, Tustin, CA, 1995.

Hodge, D.S. and Devinny, J.S., Biofilter treatment of ethanol vapors, *Environ. Prog.*, 13(3), 167, 1994.

Hodge, D.S. and Devinny, J.S., Determination of transfer rate constants and partition coefficients for air phase biofilters, *J. Environ. Eng.*, 123(6), 577, 1997.

Hodge, D.S. and Devinny, J.S., Modeling removal of air contaminants by biofiltration, *J. Environ. Eng.*, 121(1), 21, 1995.

Hugler, W.C., Cantu-De la Garza, J.G., and Villa-Garcia, M., Biofilm analysis for an odor-removing trickling filter, in *Proceedings of the 89th Annual Meeting and Exhibition of the Air and Waste Management Association*, Air and Waste Management Association, Pittsburgh, PA, 1996.

Islander, R.L., Devinny, J.S., Mansfeld, F., Postyn, A., and Shih, H., Microbial ecology of crown corrosion in sewers, *J. Environ. Eng.*, 117(6), 751, 1991.

Jennings, P.A., Snoeyink, V.L., and Chian, E.S.K., Theoretical model for a submerged biological filter, *Biotechnol. Bioeng.*, 18, 1249, 1976.

Johnson, C.T. and Deshusses, M.A., Quantitative structure-activity relationships for VOC biodegradation in biofilters, in *Proceedings of the Fourth International In Situ and On-Site Bioreclamation Symposium*, Vol. 5, Battelle Press, Columbus, OH, 1997, p. 175.

Jol, A. and Dragt, A., Filtering out volatile organic compounds from waste gases, *Process Eng.*, 9, 65, 1988.

Kamarthi, R. and Willingham, R.T., Bench-scale evaluation of air pollution control technology based on biological treatment process, in *Proceedings of the 87th Annual Meeting and Exhibition of the Air and Waste Management Association*, Air and Waste Management Association, Pittsburgh, PA, 1994.

Kampeter, S.M., Personal communication, Monsanto Enviro-Chem, St. Louis, MO, 1998.

Kier, L.B. and Hall, L.H., *Molecular Connectivity in Structure-Activity Analysis*, John Wiley & Sons, New York, 1986.

Kinney, K., du Plessis, C., Schroeder, E.D., Chang, D.P.Y.and Scow, K.M., Optimizing microbial activity in a directionally switching biofilter, in *Proceedings of the 1996 Conference on Biofiltration (an Air Pollution Control Technology)*, Reynolds, F.E., Ed., The Reynolds Group, Tustin, California, 1996a, 150.

Kinney, K.A., Chang, D.P.Y., Schroeder, E.D., and Scow, K.M., Performance of a directionally-switching biofilter treating toluene contaminated air, in *Proceedings of the 89th Annual Meeting and Exhibition of the Air and Waste Management Association*, Air and Waste Management Association, Pittsburgh, PA, 1996b.

Kok, H.J.G., Biowassersyteem voor de behandeling van koolwaterstof afgassen, TNO report 91-151/R.22/JPO, 1991.

Kok, H.J.G., Bioscrubbing of air contaminated with high concentrations of hydrocarbons, in *Biotechniques for Air Pollution Abatement and Odour Control Policies*, Dragt, A.J. and van Ham, J., Eds., Elsevier Science, Amsterdam, 1992, p. 77.

Kosteltz, A.M., Finkelstein, A., and Sears, G., What are the "real opportunities" in biological gas cleaning for North America, in *Proceedings of the 89th Annual Meeting and Exhibition of the Air and Waste Management Association*, Air and Waste Management Association, Pittsburgh, PA, 1996.

Lackey, L.W. and Boles, J.L., Biofiltration of trichloroethylene-contaminated air streams using a propane-oxidizing consortium, in *Proceedings of the Fourth International In Situ and On-Site Bioremediation Symposium*, Vol. 5, Battelle Press, Columbus, OH, 1997, p. 189.

Lee, B.D., Apel, W.A., Walton, M.R., and Cook, L.L., Treatment of methanol contaminated air streams using biofiltration, in *Proceedings of the 89th Annual Meeting and Exhibition of the Air and Waste Management Association*, Air and Waste Management Association, Pittsburgh, PA, 1996.

Lee, B.D., Apel, W.A., and Walton, M.R., Utilization of toxic gases and vapors as alternate electron acceptors in biofilters, in *Proceedings of the 90th Annual Meeting and Exhibition of the Air and Waste Management Association*, Air and Waste Management Association, Pittsburgh, PA, 1997.

Lesley, M.P. and Chakravarthi, R.R., Integrating biofiltration with SVE: a case study, *J. Soil Contamination*, 6(1), 95, 1997.

Leson, G., Biofiltration: Investigation of a Novel Air Pollution Control Technology for the South Coast Air Basin, Ph.D. thesis, University of California, Los Angeles, 1993.

Leson, G. and Winer, A.M., Biofiltration: an innovative air pollution control technology for VOC emissions, *J. Air Waste Manage. Assoc.*, 41(8), 1045, 1991.

Leson, G. and Smith, B.J., Results from the PERF field study on biofilters for removal of volatile petroleum hydrocarbons, in *Proceedings of the 1995 Conference on Biofiltration (an Air Pollution Control Technology)*, Hodge, D.S. and Reynolds, F.E., Eds., The Reynolds Group, Tustin, CA, 1995, p. 99.

Leson, G. and Smith.B.J., Petroleum Environmental Research Forum field study on biofilters for control of volatile hydrocarbons, *J. Environ. Eng.*, 123(6), 556, 1997.

Leson, G., Hodge, D.S., Tabatabai, F., and Winer, A.M., Biofilter demonstration projects for the control of ethanol emissions, in *Proceedings of the 86th Annual Meeting and Exhibition of the Air and Waste Management Association*, Air and Waste Management Association, Pittsburgh, PA, 1993.

Leson, G., Chavira, R., Winer, A.M., and Hodge, D.S., Experiences with a full-scale biofilter for control of ethanol emissions, in *Proceedings of the 88th Annual Meeting and Exhibition of the Air and Waste Management Association*, Air and Waste Management Association, Pittsburgh, PA, 1995.

Li, D.X., *In situ* biofiltration for treatment of petroleum hydrocarbons, in *Proceedings of the 1995 Conference on Biofiltration (an Air Pollution Control Technology)*, Hodge, D.S. and Reynolds, F.E., Eds., The Reynolds Group, Tustin, CA, 1995, p. 1.

Lipski, A., Droege, A., Reichart, K., and Altendorf, K.H., Detection and diversity of styrene degrading micro-organism from biofilters, in *Biological Waste Gas Cleaning, Proceedings of an International Symposium*, Prins, W.L. and van Ham, J., Eds., VDI Verlag GmbH, Duesseldorf, Germany, 1997, p. 265.

Liu, P.K.T., Gregg, R.L., Sabol, H.K., and Barkley, N., Engineered biofilter for removing organic contaminants in air, *J. Air Waste Manage. Assoc.*, 44(3), 299, 1994.

Loy, J., Personal communication, Zander Umwelt GmbH, Nuremberg, Germany, 1998.

Loy, J., Heinrich, K., and Egerer, B., Influence of filter material on the elimination rate in a biotrickling filter bed, in *Proceedings of the 90th Annual Meeting and Exhibition of the Air and Waste Management Association*, Air and Waste Management Association, Pittsburgh, PA, 1997.

Medina, V.F., Devinny, J.S., and Ramaratnam, M., Biofiltration of toluene vapors in a carbon-medium biofilter, in *Biological Unit Processes for Hazardous Waste Treatment, Proceedings of the Third International In Situ and On-Site Bioreclamation Symposium*, Hinchee, R.E., Skeen, R.S., and Sayles, G.D., Eds., Battelle Press, Columbus, OH, 1995a, p. 257.

Medina, V.F., Webster, T., Ramaratnam, M., and Devinny, J.S., Treatment of gasoline residuals by granular activated carbon based biological filtration, *J. Environ. Sci. Health*, 30(2), 407, 1995b.

Menig, H., Krill, H., and Jaeschke, T., Kosten und Effizienz biologischer Verfahren im Vergleich zu anderen Abluftreinigungssystemen, in *Biological Waste Gas Cleaning, Proceedings of an International Symposium*, Prins, W.L. and van Ham, J., Eds., VDI Verlag GmbH, Duesseldorf, Germany, 1997, p. 27.

Mildenberger, H.J. and van Lith, C., Long-term experiences with large enclosed biofilter at sewage water treatment facility, in *Proceedings of the 89th Annual Meeting and Exhibition of the Air and Waste Management Association*, Air and Waste Management Association, Pittsburgh, PA, 1996.

Mirpuri, R., Sharp, W., Villaverde, S., Jones, W., Lewandowski, Z., and Cunningham, A., Predictive model for toluene degradation and microbial phenotypic profiles in flat plate vapor phase bioreactor, *J. Environ. Eng.*, 123(6), 586, 1997

Moe, W.L. and Irvine, R.L., Polyurethane foam based biofilter media for toluene removal, presented at Emerging Technologies in Hazardous Waste Managment IX, Special Symposium, Industrial and Engineering Chemistry Division, American Chemical Society, Pittsburgh, PA, 1997.

Mohseni, M. and Allen, D.G., Biofiltration of alpha-pinene using wood waste and activated carbon media, in *Proceedings of the 1996 Conference on Biofiltration (an Air Pollution Control Technology)*, Reynolds, F.E., Ed., The Reynolds Group, Tustin, CA, 1996, p. 45.

Møller, S., Pedersen, A.R., Poulsen, L.K., Arvin, E., and Molin, S., Activity and three-dimensional distribution of toluene-degrading *Pseudomonas putida* in a multispecies biofilm assessed by quantitative in situ hybridization and scanning confocal laser microscopy, *Appl. Environ. Microbiol.*, 62(12), 4632, 1996.

Morgenroth, E.E., Schroeder, E.D., Chang, D.P.Y., and Scow, K.M., Nutrient limitation in a compost biofilter degrading hexane, in *Proceedings of the 88th Annual Meeting and Exhibition of the Air and Waste Management Association*, Air and Waste Management Association, Pittsburgh, 1995a.

Morgenroth, E., Schroeder, E.D., Chang, D.P.Y., and Scow, K.M., Modeling of a compost biofilter incorporating microbial growth, in *Innovative Technologies for On-Site Remediation and Hazardous Waste Management, Proceedings of the National Conference of The Environmental Engineering Division, J. Air Waste Manage. Assoc.*, American Society of Civil Engineers, New York, 1995b, p. 473.

Morgenroth, E., Schroeder, E.D., Chang, D.P.Y., and Scow, K.M., Nutrient limitation in a compost biofilter degrading hexane, *J. Air Waste Manage. Assoc.*, 46(4), 300, 1996.

Morton, R.L. and Caballero, R.C., The biotrickling story: LA County Sanitation Districts' research pays off, *Water Environ. Technol.*, 8(6), 39, 1996.

Morton, R.L. and Caballero, R.C., Removing hydrogen sulfide from wastewater treatment facilities' air process streams with a biotrickling filter, in *Proceedings of the 90th Annual Meeting and Exhibition of the Air and Waste Management Association*, Air and Waste Management Association, Pittsburgh, PA, 1997.

Neilson, L.M., Kim, B.J., and Severin, B.F., Eucaryotic monitoring of the exhaust from biofilters, in *Proceedings of the 1995 Conference on Biofiltration (an Air Pollution Control Technology)*, Hodge, D.S. and Reynolds, F.E., Eds., The Reynolds Group, Tustin, CA, 1995, p. 247.

Norris, R.D. et al., *Handbook of Bioremediation*, Robert S. Kerr Environmental Laboratory and CRC Press, Boca Raton, FL, 1994.

Okey, R.W. and Stensel, D., A QSAR-based biodegradability model — a QSBR, *Water Res.*, 30(9), 2206, 1996.

Ortíz, I., Morales, M., Gobbée, C., Guerrero, V., Auria, R., and Revah, S., Biofiltration of gasoline VOCs with different support media, in *Proceedings of the 91st Annual Meeting and Exhibition of the Air and Waste Management Association*, Air and Waste Management Association, Pittsburgh, PA, 1998.

Ottengraf, S.P.P., Theoretical model for a submerged biological filter, *Biotechnol. Bioeng.*, 19, 1411, 1977.

Ottengraf, S.P.P., Exhaust gas purification, in *Biotechnology*, Vol. 8, Rehm, H.J. and Reed, G., Eds., VCH Verlagsgesellschaft, Weinheim, 1986.

Ottengraf, S.P.P., *Biological Systems for Waste Gas Elimination*, Elsevier Science, Amsterdam, 1987.

Ottengraf, S.P.P. and van den Oever, A.H.C., Kinetics of organic compound removal from waste gases with a biological filter, *Biotechnol. Bioeng.*, 25(12), 3089, 1983.

Ottengraf, S.P.P. and Konings, J.H.G., Emission of microorganisms from biofilters, *Bioprocess Eng.*, 7(1–2), 89, 1991.

Ottengraf, S.P.P. and Diks, R.M.M., Process technology of biotechniques, in *Biotechniques for Air Pollution Abatement and Odour Control Policies*, Dragt, A.J. and van Ham, J., Eds., Elsevier Science, Amsterdam, 1992, p. 17.

Ottengraf, S.P.P., Meesters, J.J., van den Oever, A.H.C., and Rozema, H.R., Biological elimination of volatile xenobiotic compounds in biofilters, *Bioprocess Eng.*, 1, 61, 1986.

Oude Luttighuis, H.H.F., A new generation of packing materials for biofilters, in *Biological Waste Gas Cleaning, Proceedings of an International Symposium*, Prins, W.L. and van Ham, J., Eds., VDI Verlag GmbH, Duesseldorf, Germany, 1997, p. 141.

Paca, J., Marek, J., Weigner, P., and Koutsky, B., Biofilter characteristics at high xylene and toluene loadings, in *Biological Waste Gas Cleaning*, Prins, W.L. and van Ham, J., Eds., VDI Verlag GmbH, Dusseldorf, Germany, 1997, p. 123.

Paul, H., Biofiltereinsatz zur Reduzierung der Geruchsemissions aus einem Giessereibetrieb, in *Biologische Abgasreinigung*, VDI Berichte 1104, VDI Verlag GmbH, Duesseldorf, Germany, 1994, p. 355.

Paul, H. and Sabo, F., Reduction of odorous emission from foundries using biofilters, in *Proceedings of the 1995 Conference on Biofiltration (an Air Pollution Control Technology)*, Hodge, D.S. and Reynolds, F.E., Eds., The Reynolds Group, Tustin, CA, 1995, p. 333.

Paul, H. and Nisi, D., Four years experience with biofilter technology in the automotive, in *Proceedings of the 1996 Conference on Biofiltration (an Air Pollution Control Technology)*, Reynolds, F.E., Ed., The Reynolds Group, Tustin, CA, 1996, p. 55.

Pedersen, A.R., Moller, S., Molin, S., and Arvin, E., Activity of toluene-degrading *Pseudomonas putida* in the early growth phase of a biofilm for waste gas treatment, *Biotechnol. Bioeng.*, 54(2), 131, 1997.

Pinnette, J.R., Dwinal, C.A., Giggey, M.D., and Hendry, G.E., Design implications of the biofilter heat and moisture balance, in *Proceedings of the 1995 Conference on Biofiltration (an Air Pollution Control Technology)*, Hodge, D.S. and Reynolds, F.E., Eds., The Reynolds Group, Tustin, CA, 1995, p. 85.

Pomeroy, R.D., De-odorizing of Gas Streams by the Use of Microbial Growths, U.S. Patent 2,793,096, 1957.

Potera, C., Research news: biofilms invade microbiology, *Science*, 273, 1795, 1996.

Proell, F., Friedl, A., and Zich, T., Pilot-Versuche bei Industriebetrieben mit Kompostpellets, einem neuartigen Biofilterfuellmaterial, in *Biological Waste Gas Cleaning, Proceedings of an International Symposium*, Prins, W.L. and van Ham, J., Eds., VDI Verlag GmbH, Duesseldorf, Germany, 1997, p. 173.

Sabo, F., Behandlung von Deponiegas im Biofilter, Ph.D. thesis, University of Stuttgart, Germany, 1991.

Sabo, F., Schneider, T., and Motz, U., Latest developments and industrial applications of biofiltration, in *Proceedings of 1996 Conference on Biofiltration (an Air Pollution Control Technology)*, Reynolds, F.E., Ed., Reynolds Group, Tustin, CA, 1996, p. 63.

Seed, L.P. and Corsi, R.L., Biofiltration of BTEX contaminated streams: laboratory studies, in *Proceedings of 87th Annual Meeting and Exhibition of the Air and Waste Management Association*, Air and Waste Management Assoc., Pittsburgh, PA, 1994.

Seed, L.P. and Corsi, R.L., Biofiltration of benzene, toluene, and o-xylene: substrate effects and carbon balancing, in *Proceedings of the 89th Annual Meeting and Exhibition of the Air and Waste Management Association*, Air and Waste Management Association, Pittsburgh, PA, 1996.

Shareefdeen, Z. and Baltzis, B.C., Biofiltration of toluene vapor under steady-state and transient conditions: theory and experimental results, *Chem. Eng. Sci.*, 49, 4347, 1994.

Shareefdeen, Z., Baltzis, B.C., Oh, Y.S., and Bartha, R., Biofiltration of methanol vapor, *Biotechnol. Bioeng.*, 41, 512, 1993.

Siegrist, H. and Gujer, W., Mass transfer mechanisms in a heterotrophic biofilm, *Water Res.*, 19(11), 1369, 1985.

Skladany, G.J., Deshusses, M.A., Devinny, J.S., Togna, A.P., and Webster, T.S., Biofilters, in *Handbook of Odor and VOC Control*, McGraw-Hill, New York, 1998.

Sly, L.I., Bryant, L.J., Cox, J.M., and Anderson, J.M., Development of a biofilter for the removal of methane from coal mine ventilation atmospheres, *Appl. Microbiol. Biotechnol.*, 39, 400, 1993.

Smet, E., van Langehove, H., and Verstraete, W., Long-term stability of a biofilter treating dimethyl sulphide, *Appl. Microbiol. Biotechnol.*, 46(2), 191, 1996.

Smith, F.L., Sorial, G.A., Suidan, M.T., Breen, A.W., and Biswas, P., Development of two biomass control strategies for extended stable operation of highly efficient biofilters with high toluene loadings, *Environ. Sci. Technol.*, 30(5), 1744, 1996.

Sorial, G.A., Smith, F.L., Suidan, M.T., Biswas, P., and Brenner, R.C., Evaluation of a trickle bed biofilter for toluene removal, *J. Air Waste Manage. Assoc.*, 45(12), 801, 1995.

Sorial, G.A., Smith, F.L., Suidan, M.T., Pandit, A., Biswas, P., and Brenner, R.C., Evaluation of trickle bed air biofilter performance for BTEX removal, *J. Environ. Eng.*, 123(6), 530, 1997.

Speitel, G.E. and McLay, D.S., Biofilm reactors for treatment of gas streams containing chlorinated solvents, *J. Environ. Eng.*, 119(4), 658, 1993.

Standefer, S., Personal communication, PPC Biofiltration, Longview, TX, 1998.

Standefer, S. and Willingham, R., Experience with pilot and full-scale biofilter operations, in *Proceedings of the 1996 Conference on Biofiltration (an Air Pollution Control Technology)*, Reynolds, F.E., Ed., The Reynolds Group, Tustin, CA, 1996, p. 77.

Stewart, W.C. and Thom, R.R., Test results and economics of using an innovative, high-rate, vapor-phase biofilter in industrial applications, in *Proceedings of the 89th Annual Meeting and Exhibition of the Air and Waste Management Association*, Air and Waste Management Association, Pittsburgh, PA, 1996.

Stewart, W.C. and Kamarthi, R.S., Biofilter application for control of BTEX compounds from glycol dehydrator condenser vent gases at oil and natural gas producing facility, in *Proceedings of the 90th Annual Meeting and Exhibition of the Air and Waste Management Association*, Air and Waste Management Association, Pittsburgh, PA, 1997.

Stolp, H., *Microbial Ecology: Organisms, Habitats, Activities*, Cambridge University Press, London, 1988.

Sukesan, S. and Watwood, M.E., Removal of trichloroethylene in compost-packed biofiltration columns, in *Proceedings of the Fourth International In Situ and On-Site Bioremediation Symposium*, Vol. 5, Battelle Press, Columbus, OH, 1997, p. 183.

Swanson, W.J. and Loehr, R.C., Biofiltration: fundamentals, design and operations principles, and applications, *J. Environ. Eng.*, 123(6), 538, 1997.

Thoenes, D., Jr., and Kramers, H., Mass transfer from spheres in various regular packings to a flowing fluid, *Chem. Eng. Sci.*, 8, 271, 1958.

Thompson, D., Sterne, L., Bell, J., Parker, W., and Lye, A., Pilot scale investigation of sustainable BTEX removal with a compost biofilter, in *Proceedings of the 89th Annual Meeting and Exhibition of the Air and Waste Management Association*, Air and Waste Management Association, Pittsburgh, PA, 1996.

Togna, A.P. and Folsom, B.R., Removal of styrene from air using bench-scale biofilter and biotrickling filter reactors, in *Proceedings of the 85th Annual Meeting and Exhibition of the Air and Waste Management Association*, Air and Waste Management Association, Pittsburgh, PA, 1992.

Togna, A.P., Skladany, G.J., and Caratura, J.M., Treatment of BTEX and petroleum hydrocarbon vapors using a field-pilot biofilter, in *Proceeding of the 49th Annual Purdue University Industrial Waste Conference*, Purdue University, W. Lafayette, IN, 1994.

Togna, A.P., Fucich, W.J., Loudon, R.E., Del Vecchio, M., Barshter, D.W., and Nadeau, A.J., Treatment of odorous toxic pollutants from a hardwood panel board manufacturing facility using biofiltration, in *Proceedings of the 90th Annual Meeting and Exhibition of the Air and Waste Management Association*, Air and Waste Management Association, Pittsburgh, PA, 1997.

Unger, D.R., Lam, T.T., Schaefer, C.E., and Kosson, D.S., Predicting the effect of moisture on vapor-phase sorption of volatile organic compounds to soils, *Environ. Sci. Technol.*, 30(4), 1081, 1996.

van Groenestijn, J., Harkes, M., Cox, H., and Doddema, H., Ceramic materials in biofiltration, in *Proceedings of the 1995 Conference on Biofiltration (an Air Pollution Control Technology)*, Hodge, D.S. and Reynolds, F.E., Eds., The Reynolds Group, Tustin, CA, 1995, p. 317.

van Langenhove, H. and Smet, E., Biofiltration of organic sulfur compounds, in *Proceedings of the 1996 Conference on Biofiltration (an Air Pollution Control Technology)*, Reynolds, F.E., Ed., The Reynolds Group, Tustin, CA, 1996, p. 206.

van Langenhove, H., Lootens, A., and Schamp, N., Inhibitory effects of SO_2 on biofiltration of aldehydes, *Water Air Soil Pollution*, 47(1–2), 81, 1989.

van Lith, C., Design criteria for biofilters, in *Proceedings of the 82nd Annual Meeting and Exhibition of the Air and Waste Management Association*, Air and Waste Management Association, Pittsburgh, PA, 1989.

van Lith, C., David, S.L., and Marsh, R., Design criteria for biofilters, *Trans. I. Chem. E.*, 68, 127, 1990.

van Lith, C., Leson, G., and Michelson, R., Evaluating design options for biofilters, in *Proceedings of the 1996 Conference on Biofiltration (an Air Pollution Control Technology)*, Reynolds, F.E., Ed., The Reynolds Group, Tustin, CA, 1996, p. 102.

van Lith, C., Leson, G., and Michelsen, R., Evaluating design options for biofilters, *J. Air Waste Manage. Assoc.*, 47(1), 37, 1997.

VDI-Berichte 3477, Biological waste gas/waste air purification: biofilters, in *VDI-Handbuch Reinhalten der Luft*, Band 6, Dusseldorf, 1991, p. 32.

Ventera, R. and Findlay, M., Biofiltration for the treatment of airstreams contaminated with jet fuel vapors, in *Proceedings of the New England Environmental Exposition*, Longwood Environmental Management, Boston, 1991.

Veir, J.K., Schroeder, E.D., Chang, D.P.Y., and Scow, K.M., Interaction between toluene and dichloromethane degrading populations in a compost biofilter, in *Proceedings of the 89th Annual Meeting and Exhibition of the Air and Waste Management Association*, Air and Waste Management Association, Pittsburgh, PA, 1996.

Vestal, J.R. and White, D.C., Lipid analysis in microbial ecology, *Bioscience*, 39(8), 535, 1989.

Weber, F.J. and Hartmans, S., Use of activated carbon as a buffer in biofiltration of waste gases with fluctuating concentrations of toluene, *Appl. Microbiol. Biotechnol.*, 43(2), 365, 1995.

Weber, F.J. and Hartmans, S., Prevention of clogging in a biological trickle-bed reactor removing toluene from contaminated air, *Biotechnol. Bioeng.*, 50(1), 91, 1996.

Webster, N., Personal communication, Webster Environmental Associates, Pewee Valley, KT, 1998.

Webster, T.S., Control of Air Emissions from Publicly Owned Treatment Works Using Biological Filtration, Ph.D. thesis, The University of Southern California, Los Angeles, 1996.

Webster, T.S. and Devinny, J.S., Biofiltration technology for air pollution control, in *Encyclopedia of Environmental Analysis*, Meyers, R.A., Ed., John Wiley & Sons, New York, 1998.

Webster, T.S., Devinny, J.S., Torres, E.M., and Basrai, S., Study of biofiltration for control of odor, VOC and toxic air emissions from wastewater treatment plants — phase II bench- and pilot-scale studies, in *Proceedings of the 1995 Conference on Biofiltration (an Air Pollution Control Technology)*, Hodge, D.S. and Reynolds, F.E., Eds., The Reynolds Group, Tustin, CA, 1995, p. 259.

Webster, T.S., Devinny, J.S., Torres, E.M., and Basrai, S.S., Biofiltration of odors, toxics and volatile organic compounds from publicly owned treatment works, *Environ. Prog.*, 15(3), 141, 1996.

Webster, T.S., Devinny, J.S., Torres, E.M., and Basrai, S.S., Microbial ecosystems in compost and granular activated carbon biofilters, *Biotechnol. Bioeng.*, 53(3), 296, 1997.

White, D.C., Ringelberg, D.B., Pfiffner, S.M., Pinkart, H.C., Nivens, D.E., and Lane, J., Utility of signature lipid biomarker analysis of the extant microbiota on monitoring *in situ* bioremediation effectiveness, in *Proceedings of the Third International In Situ and On-Site Bioremediation Symposium*, Battelle Press, Columbus, OH, 1995.

Williams, T.O. and Boyette, R.A., Biofiltration for odor control at a sewage interceptor pumping station in Charlottesville, Virginia, in *Proceedings of the 1995 Conference on Biofiltration (an Air Pollution Control Technology)*, Hodge, D.S. and Reynolds, F.E., Eds., The Reynolds Group, Tustin, CA, 1995, p. 303.

Williamson, K. and McCarty, P.L., A model of substrate utilization by bacterial films, *J. Water Pollution Control Fed.*, 48(2), 9, 1976.

Wolstenholme, P. and Finger, R., Long-term odor and VOC pilot tests on biofilters, in *Proceedings of the 1995 Conference on Biofiltration (an Air Pollution Control Technology)*, Hodge, D.S. and Reynolds, F.E., Eds., The Reynolds Group, Tustin, CA, 1995, p. 273.

Wright, W.F., Schroeder, E.D., Chang, D.P.Y., and Romstad, K., Performance of a pilot-scale compost biofilter treating gasoline vapor, *J. Environ. Eng.*, 123(6), 547, 1997.

Yang, Y. and Allen, E.R., Biofiltration control of hydrogen sulfide 1: design and operational parameters, *J. Air Waste Manage. Assoc.*, 44(7), 863, 1994.

Yavorsky, J., Odor and VOC control from flavor manufacturing through advanced biofiltration, in *Proceedings of the 90th Annual Meeting and Exhibition of the Air and Waste Management Association*, Air and Waste Management Association, Pittsburgh, PA, 1997.

Yudelson, J.M., The future of the U.S. biofiltration industry, in *Proceedings of the 1996 Conference on Biofiltration (an Air Pollution Control Technology)*, Reynolds, F.E., Ed., The Reynolds Group, Tustin, CA, 1996, p. 1.

Zahodiakin, P., Puzzling out the new clean air act, *Chem. Eng.*, 97(12), 24, 1995.

Zarook, S.M., Shaikh, A.A., Ansar, Z., and Baltzis, B.C., Biofiltration of volatile organic compound (VOC) mixtures under transient conditions, *Chem. Eng. Sci.*, 52(21–22), 4135, 1997.

Index

A

absorption, 4, 7, 33
acclimation, 87–89, 96, 228, 250
acetaldehyde, 93
acetic acid, 38, 63, 93, 239
activated carbon. *See* carbon:
   activated
adsorption, 7, 28–29, 57
  capacity of medium, 30, 196
  effects of, on biodegradation, 31
Aerobacter aerogenes, 56
aerobic degradation, 36
aerobic metabolism, 91
air distribution system, 161–164,
   183, 237
air flow, 190–194, 218
  direction, 71–72, 101
  rates, 188–189, 195, 196, 204
air load, 190–195
air pollution legislation, 1, 2–3
air toxics, 2
air-phase reactor, 7, 96
alcohols, 5, 16, 162, 223, 231, 234, 247
aldehydes, 5, 87, 135–136, 144, 162
algae, 98
aliphatic compounds, 17, 212, 226
alkaline buffers, 160
alkalinity, 62–65, 200–201
  defined, 64
alkanes, 63, 228

American Chemical Society, 21
ammonia, 5, 16, 36, 37, 66
amoebae, 98
anaerobic metabolism, 91
anaerobic sludge digesters, 63
analytical systems for biofilters, 173
applications, 211–250
  ARA-Rhein biofilter, 211–214
  City of Poughkeepsie biofilter,
   218–220
  ethanol from a foundry off-gas,
   235
  exhaust air in wood industry,
   230–232
  fabric softener facility, 220–224,
  flavor and fragrance
   manufacturing, 215
  flavor manufacturing, 216–218
  flexographic printing off-gases,
   245–247
  gasoline vapor treatment, 225–230
  high-concentration, low-flow
   biofilter, 243–245
  ink-drying operations, 232–235
  removal of VOCs and odors,
   240–243
  tobacco company biotrickling
   filter, 248–250
aromatics, 5, 17, 87, 135–136, 144,
   162, 227, 240
autoclaving, 33

289

B

backflushing, 173
backwashing, 101
bacteria, 82, 90, 98, 99
   density of, 98
   heterotrophic, 101
   population in biofilters, 107
   vs. fungi, 82, 99
bacterial attachment of media, 44
baffles, 186, 204
base, amount of, 65
bed exhaustion, 4
bench-scale biofilter, typical, 147
bench-scale testing, 144–149
benzene, 17, 87, 89, 226, 255
benzoic acid, 72
bicarbonate wash, 239
biodegradation of contaminants,
      31–32, 34–37. *See also*
   contaminants: biodegradation
   of
biofilms, 34, 94–99
   models of, 113–134
biofiltration, 12
   air flow, 26
      direction, 71–72, 101
      interference in, 58
   cometabolism, 76–77. *See also*
      cometabolism
   computer control of, 172–174. *See
      also* programmable logic
      controller (PLC)
   controlling factors of, 51–79
   cost-effective conditions for, 13
   costs, 174–183. *See also* costs
   description of ,1–2, 7
   design of biofilters, 141–184. *See
      also* design of biofilters
   drainage, 59
   extreme, 73–79
   heat generated by, 39–40
   high-temperature, 78
   internal mechanisms of, 8, 23–40
   low-water-content type, 78–79

marketplace, 15
mechanisms of, 23–40
media, 41. *See also* media
   criteria for, 42–45
   description of, 49–50
   materials used for, 45–49. *See
      also* individual types of
      materials
microbial ecology of, 81–110
modeling of, 111–139
nutrients, 65–67
operation of filters, 51–79
oxygen limitation, 68, 239
product generation by, 37–39
review of, 13–15
temperature, 60–62. *See also*
   temperature
terminology of, 17–21
thermodynamic relationships,
   55–56
typical operating conditions, 10
vs. biotrickling filter, 248–250
water content, 51–60
   control of, 59–60
biological activity, 205–207
   cell counts, 205–206
   fatty acid analysis, 207
   respiration, 206
   visual inspection, 207
biological effects, in biofilter, 56–57
biomass
   accumulation, 38
   and biological activity, 206–207
   clogging, 100–102
   growth, limiting, 38
biopolymers, 33
bioreactors, classification of, 7
bioscrubbers, 8–10, 11
biosurfactants, 32
BIOTON®, 42, 50
biotransformation, 36
biotrickling filters, 8–10, 12, 28, 56,
   65, 85, 90, 94, 97, 98, 248–250
biowasher, 11
biphenyl, 32

Milton Keynes UK
Ingram Content Group UK Ltd.
UKHW021620071024
449327UK00020BA/1130